Geometry

Teaching Geometry with Manipulatives

 **Glencoe**

New York, New York   Columbus, Ohio   Chicago, Illinois   Peoria, Illinois   Woodland Hills, California

Manipulatives

Glencoe offers three types of kits to enhance the use of manipulatives in your Middle School Mathematics classroom.

- The **Glencoe Mathematics Overhead Manipulative Resources** contains translucent manipulatives designed for use with an overhead projector.
- The **Glencoe Mathematics Classroom Manipulative Kit** contains classroom sets of frequently used manipulatives in algebra, geometry, measurement, probability, and statistics.
- The **Glencoe Mathematics Student Manipulative Kit** contains an individual set of manipulatives often used in Student Edition activities.

The manipulatives contained in each of these kits are listed on page vi of this booklet.

Each of these kits can be ordered from Glencoe by calling (800) 334-7344.

Glencoe Mathematics Overhead Manipulative Kit	0-07-830593-4
Glencoe Mathematics Classroom Manipulative Kit	0-02-833116-8
Glencoe Mathematics Student Manipulative Kit	0-02-833654-2

The McGraw-Hill Companies

Copyright © by The McGraw-Hill Companies, Inc. All rights reserved. Permission is granted to reproduce the material contained herein on the condition that such materials be reproduced only for classroom use; be provided to students, teachers, and families without charge; and be used solely in conjunction with the *Glencoe Geometry* program. Any other reproduction, for sale or other use, is expressly prohibited.

Send all inquiries to:
Glencoe/McGraw-Hill
8787 Orion Place
Columbus, OH 43240-4027

ISBN: 0-07-860201-7   *Teaching Geometry with Manipulatives*

Printed in the United States of America.

1 2 3 4 5 6 7 8 9 10 071 12 11 10 09 08 07 06 05 04 03

Contents

Easy-to-Make Manipulatives	**Page**
Grid Paper	1
Coordinate Planes	2
Centimeter Grid Paper	3
10 × 10 Centimeter Grids	4
Coordinate Planes in Space	5
Rectangular Dot Paper	6
Isometric Dot Paper	7
Two-Column Proof Format	8
Tetrahedron Pattern	9
Hexahedron Pattern	10
Icosahedron Pattern	11
Dodecahedron Pattern	12
Octahedron Pattern	13
Tangram Pattern	14
Tessellation Shapes	15
Protractors	16
Rulers	17
Trigonometric Ratios	18
Problem Solving Guide	19

CHAPTER 1 — Page
	Teaching Notes and Overview	21
1-1	Geometry Activity Recording Sheet	26
1-1	Mini-Project: Intersecting Planes	27
1-2	Geometry Activity Recording Sheet	28
1-3	Geometry Activity Recording Sheet	29
1-3	Using Overhead Manipulatives: Bisecting a Segment	30
1-3	Geometry Activity: Pythagorean Puzzle	31
1-3	Geometry Activity Recording Sheet	34
1-4	Using Overhead Manipulatives: Measuring Angles	35
1-4	Using Overhead Manipulatives: Constructing Congruent Angles	36
1-4	Geometry Activity Recording Sheet	37
1-4	Using Overhead Manipulatives: Constructing Angle Bisectors	38
1-5	Geometry Activity Recording Sheet	39
1-5	Geometry Activity Recording Sheet	40

CHAPTER 2 — Page
	Teaching Notes and Overview	41
2-3	Geometry Activity: If-Then Statements	43
2-4	Geometry Activity Recording Sheet	46
2-4	Mini-Project: Tracing Strategy	47
2-8	Geometry Activity Recording Sheet	48

CHAPTER 3 — Page
	Teaching Notes and Overview	49
3-1	Geometry Activity Recording Sheet	52
3-2	Using Overhead Manipulatives: Parallels and Transversals	53
3-4	Mini-Project: Avoiding Lattice Points	54
3-4	Geometry Activity: Graphing Lines in the Coordinate Plane	55
3-5	Using Overhead Manipulatives: Constructing Parallel Lines	57
3-6	Using Overhead Manipulatives: Constructing Perpendicular Lines	58
3-6	Using Overhead Manipulatives: Parallels and Distance	60
3-6	Geometry Activity Recording Sheet	63

CHAPTER 4 — Page
	Teaching Notes and Overview	65
4-1	Geometry Activity Recording Sheet	69
4-2	Geometry Activity Recording Sheet	70
4-2	Using Overhead Manipulatives: Angle Measures in Triangles	71
4-3	Using Overhead Manipulatives: Congruent Triangles	72
4-3	Geometry Activity: Congruent Triangles	73
4-4	Using Overhead Manipulatives: Tests for Congruent Triangles	75
4-5	Geometry Activity Recording Sheet	77
4-5	Geometry Activity Recording Sheet	78
4-6	Geometry Activity Recording Sheet	80
4-6	Mini-Project: Perimeters and Unknown Values	81

CHAPTER 5 — Page
	Teaching Notes and Overview	83
5-1	Geometry Activity Recording Sheet	87
5-1	Using Overhead Manipulatives: Constructing a Median in a Triangle	89
5-1	Using Overhead Manipulatives: Investigating Perpendicular Bisectors	90
5-1	Geometry Activity: Special Segments in a Triangle	91
5-1	Mini-Project: Folding Triangles	93
5-2	Geometry Activity Recording Sheet	94
5-4	Using Overhead Manipulatives: Inequalities in Triangles	95
5-4	Using Overhead Manipulatives: Investigating the Triangle Inequality Theorem	96

CHAPTER 6		Page
	Teaching Notes and Overview 99	
6-3	Geometry Activity Recording Sheet 102	
6-3	Using Overhead Manipulatives: Similar Triangles 103	
6-3	Geometry Activity: Similar Triangles 104	
6-3	Mini-Project: Measuring Height 107	
6-4	Using Overhead Manipulatives: Trisecting a Segment 108	
6-6	Geometry Activity Recording Sheet 109	

CHAPTER 7		Page
	Teaching Notes and Overview 111	
7-2	Mini-Project: Pythagorean Theorem 114	
7-2	Geometry Activity Recording Sheet 115	
7-2	Using Overhead Manipulatives: The Pythagorean Theorem 116	
7-4	Geometry Activity Recording Sheet 117	
7-4	Geometry Activity: Trigonometry 118	
7-7	Geometry Activity Recording Sheet 121	

CHAPTER 8		Page
	Teaching Notes and Overview 123	
8-1	Using Overhead Manipulatives: Investigating the Exterior Angles of a Convex Polygon 127	
8-1	Geometry Activity Recording Sheet 128	
8-2	Geometry Activity Recording Sheet 129	
8-3	Using Overhead Manipulatives: Tests for Parallelograms 130	
8-3	Geometry Activity Recording Sheet 132	
8-4	Using Overhead Manipulatives: Constructing a Rectangle 133	
8-5	Using Overhead Manipulatives: Constructing a Rhombus 134	
8-5	Mini-Project: Square Search 135	
8-5	Geometry Activity Recording Sheet 136	
8-6	Geometry Activity Recording Sheet 137	
8-7	Geometry Activity: Linear Equations 138	

CHAPTER 9		Page
	Teaching Notes and Overview 141	
9-1	Geometry Activity Recording Sheet 145	
9-1	Using Overhead Manipulatives: Constructing Reflections in a Line 146	
9-2	Using Overhead Manipulatives: Translations .147	
9-2	Geometry Activity: Reflections and Translations 149	
9-2	Mini-Project: Graphing and Translations .151	
9-3	Using Overhead Manipulatives: Rotations 152	
9-4	Geometry Activity Recording Sheet 154	

9-4	Geometry Activity Recording Sheet 155	
9-6	Geometry Activity Recording Sheet 156	

CHAPTER 10		Page
	Teaching Notes and Overview 157	
10-1	Geometry Activity Recording Sheet 161	
10-3	Geometry Activity Recording Sheet 162	
10-3	Using Overhead Manipulatives: Locating the Center of a Circle 163	
10-3	Mini-Project: More About Circles 164	
10-4	Geometry Activity Recording Sheet 165	
10-4	Using Overhead Manipulatives: Investigating Inscribed Angles 166	
10-4	Geometry Activity: Inscribed Angles 167	
10-5	Using Overhead Manipulatives: Constructing a Circle to Inscribe a Triangle 170	
10-5	Using Overhead Manipulatives: Constructing Tangents 171	
10-5	Using Overhead Manipulatives: Inscribing a Circle in a Triangle 173	
10-5	Geometry Activity Recording Sheet 174	

CHAPTER 11		Page
	Teaching Notes and Overview 177	
11-1	Geometry Activity Recording Sheet 180	
11-2	Geometry Activity Recording Sheet 181	
11-2	Using Overhead Manipulatives: Investigating the Area of a Trapezoid 182	
11-3	Using Overhead Manipulatives: Constructing a Regular Hexagon 184	
11-3	Geometry Activity Recording Sheet 185	
11-3	Geometry Activity: Area of a Regular Polygon 186	
11-3	Mini-Project: Area of Circular Regions . . .189	

CHAPTER 12		Page
	Teaching Notes and Overview 191	
12-2	Using Overhead Manipulatives: Drawing a Rectangular Solid 194	
12-6	Geometry Activity: Surface Areas of Cylinders and Cones 196	
12-6	Mini-Project: Cone Patterns 199	
12-7	Geometry Activity Recording Sheet 200	
12-7	Geometry Activity Recording Sheet 201	
12-7	Using Overhead Manipulatives: Intersection of Loci 202	

CHAPTER 13		Page
	Teaching Notes and Overview 205	
13-1	Geometry Activity Recording Sheet 206	
13-2	Geometry Activity Recording Sheet 207	
13-4	Mini-Project: Word Search 208	

Teacher's Guide to Using
Teaching Geometry with Manipulatives

The book contains two sections of masters—Easy-to-Make Manipulatives and Geometry Activities. Tabs help you locate the chapter resources in each section. A complete list of manipulatives available in each of the three types of Glencoe Mathematics Manipulative Kits appears on the next page.

Easy-to-Make Manipulatives
The first section of this book contains masters for making your own manipulatives. To make more durable manipulatives, consider using card stock.

You can also make transparencies of frequently used items such as grid paper and number lines.

Activity Masters
Each chapter begins with **Teaching Notes and Overview** that summarizes the activities for the chapter and includes sample answers. There are four types of masters.

Mini-Projects are short projects that enable students to work cooperatively in small groups to investigate mathematical concepts.

Using Overhead Manipulatives provides instructions for the teacher to demonstrate an alternate approach to the concepts of the lesson by using manipulatives on the overhead projector.

Student Recording Sheets accompany the Geometry Activities found in the Student Edition. Students can easily record the results of the activity on prepared grids, charts, and figures.

Geometry Activities provide additional activities to enrich the students' experiences. These masters often include a transparency master to accompany the activity.

Glencoe Mathematics Manipulatives

Glencoe Mathematics Overhead Manipulative Resources ISBN: 0-07-830593-4	
Transparencies	**Overhead Manipulatives**
integer mat equation mat product mat inequality mat dot paper isometric dot paper coordinate grids	centimeter grid number lines lined paper regular polygons polynomial models integer models equation models

(Transparencies column and its items; Overhead Manipulatives column:)

Transparencies	Overhead Manipulatives
integer mat equation mat product mat inequality mat dot paper isometric dot paper coordinate grids	
centimeter grid number lines lined paper regular polygons polynomial models integer models equation models	algebra tiles spinners two-dimensional cups red and yellow counters decimal models (base-ten blocks) compass protractor geoboard/geobands geometric shapes transparency pens in 4 colors

Glencoe Mathematics Classroom Manipulative Kit ISBN: 0-02-833116-8		
Algebra	**Measurement, Probability, and Statistics**	**Geometry**
algebra tiles counters cups centimeter cubes equation mat/product mat coordinate grid stamp and ink pad	base-ten models marbles measuring cups number cubes protractors rulers scissors spinners stopwatches tape measures	compasses geoboards geobands geomirrors isometric dot grid stamp pattern blocks tangrams

Glencoe Mathematics Student Manipulative Kit ISBN: 0-02-833654-2	
algebra tiles red and yellow counters cups equation/product mat compass/ruler	protractor scissors geoboard geobands tape measure

Grid Paper

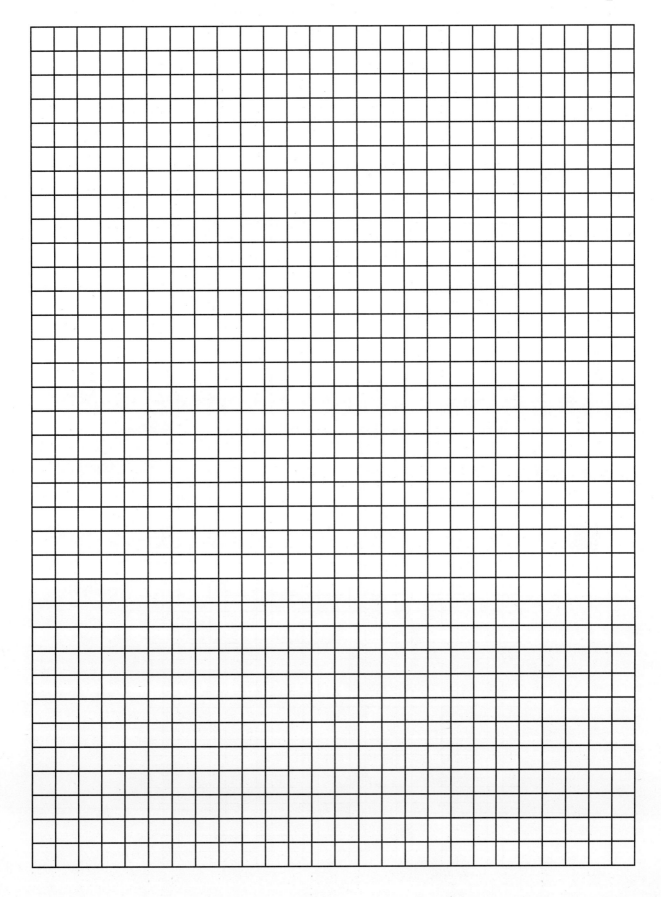

Coordinate Planes

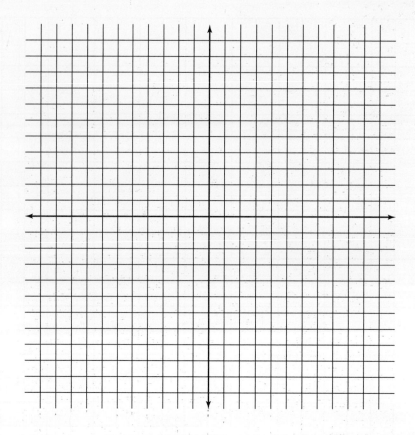

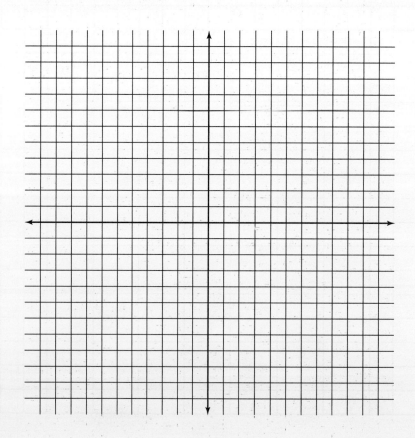

Centimeter Grid Paper

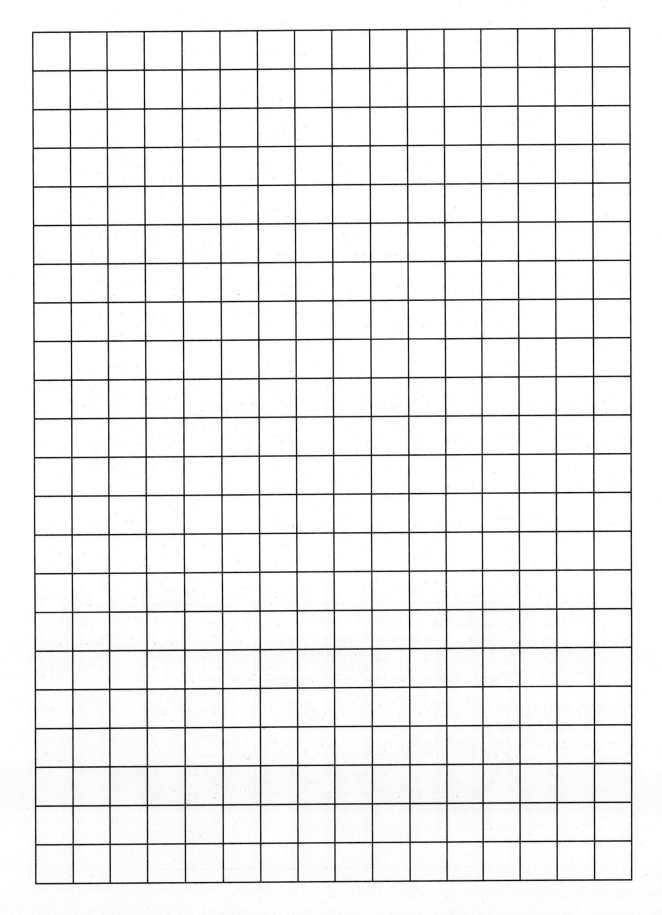

10 × 10 Centimeter Grids

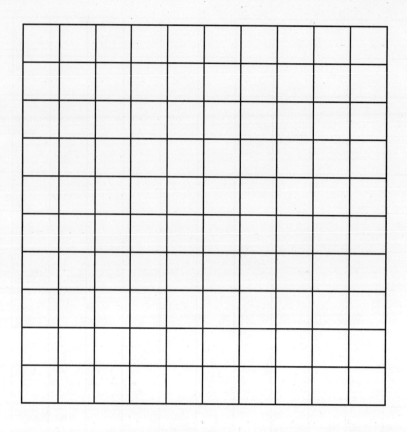

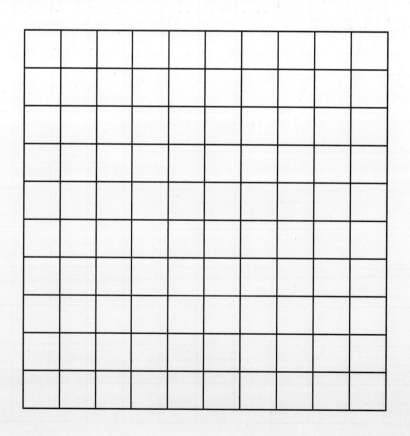

Coordinate Planes

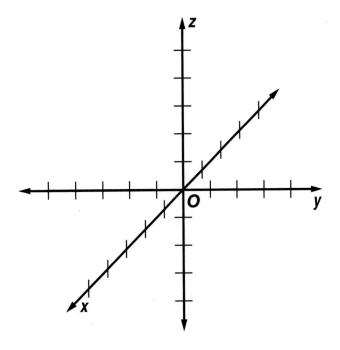

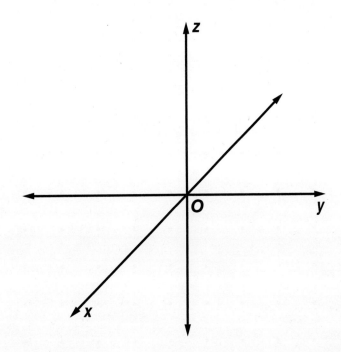

Rectangular Dot Paper

Isometric Dot Paper

Two-Column Proof Format

Given:

Figure

Prove:

Plan:

Proof:

STATEMENTS	REASONS

Tetrahedron Pattern

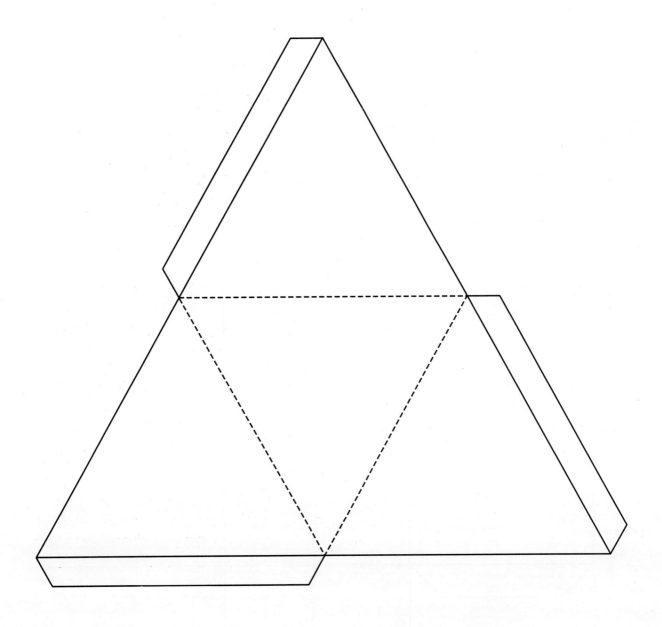

Hexahedron Pattern

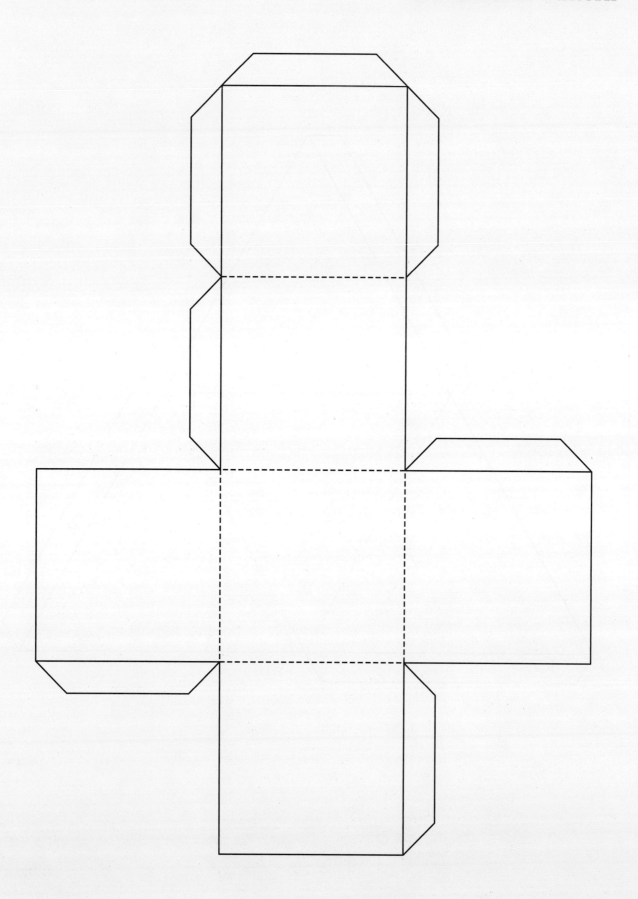

Icosahedron Pattern

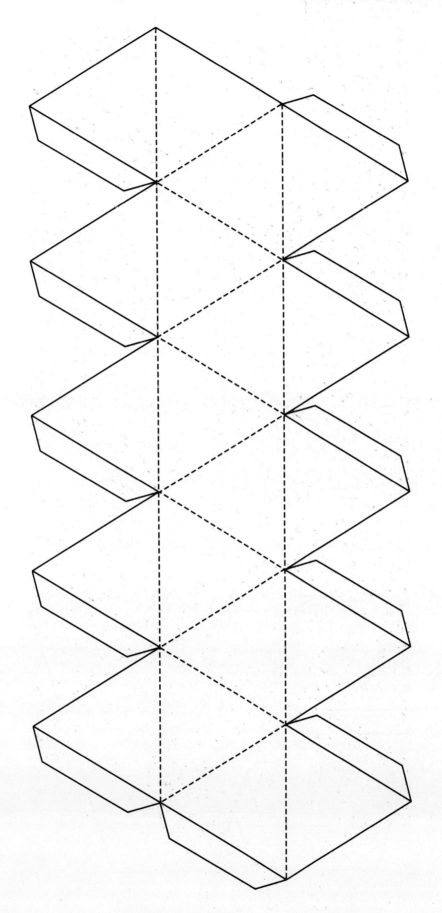

Dodecahedron Pattern

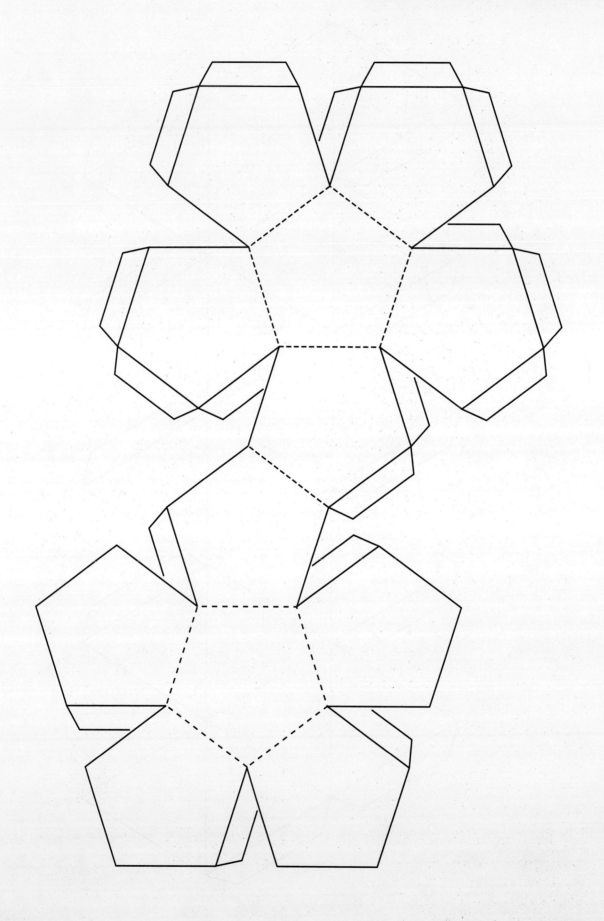

Octahedron Pattern

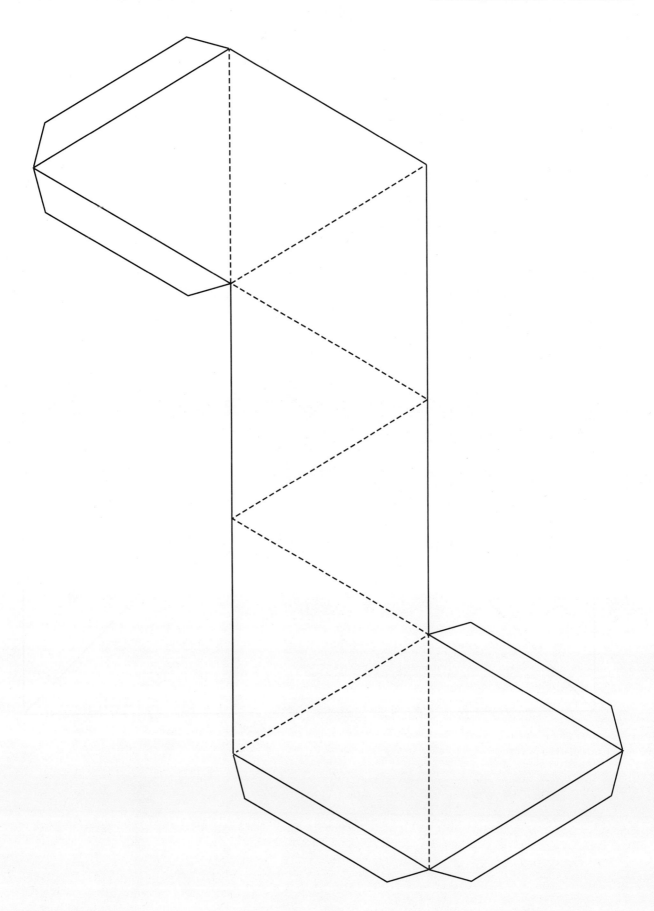

Tangram Pattern

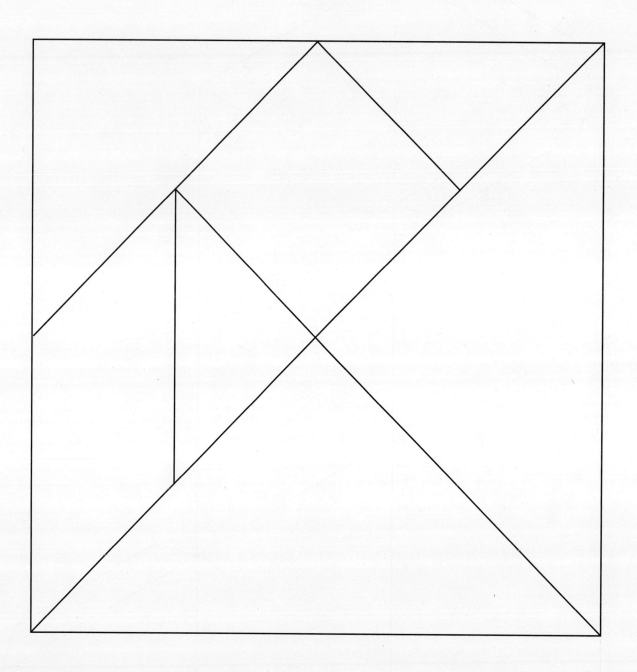

Tessellation Shapes

Protractors

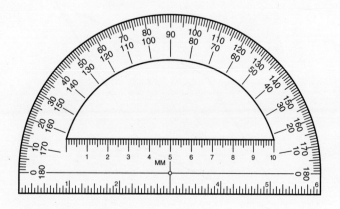
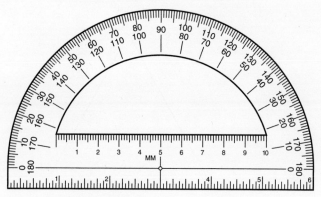
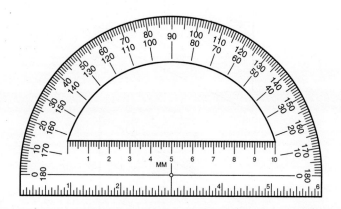
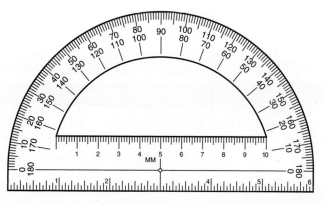
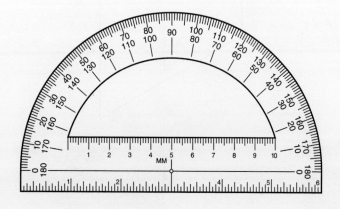
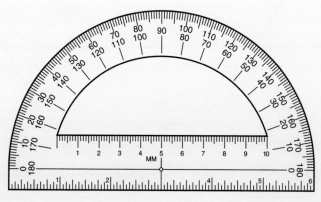

© Glencoe/McGraw-Hill — *Teaching Geometry with Manipulatives*

Rulers

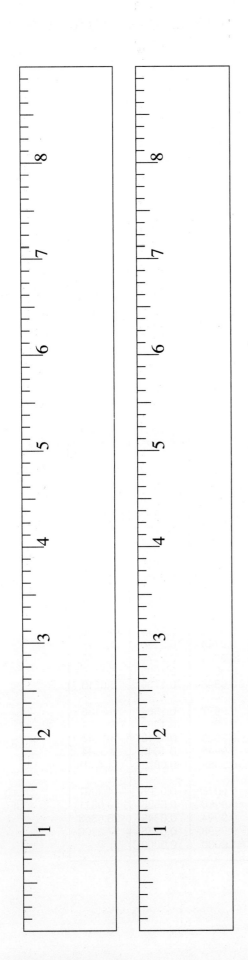

© Glencoe/McGraw-Hill **17** *Teaching Geometry with Manipulatives*

Trigonometric Ratios

TRIGONOMETRIC RATIOS

Angle	sin	cos	tan	Angle	sin	cos	tan
0°	0.0000	1.0000	0.0000	45°	0.7071	0.7071	1.0000
1°	0.0175	0.9998	0.0175	46°	0.7193	0.6947	1.0355
2°	0.0349	0.9994	0.0349	47°	0.7314	0.6820	1.0724
3°	0.0523	0.9986	0.0524	48°	0.7431	0.6691	1.1106
4°	0.0698	0.9976	0.0699	49°	0.7547	0.6561	1.1504
5°	0.0872	0.9962	0.0875	50°	0.7660	0.6428	1.1918
6°	0.1045	0.9945	0.1051	51°	0.7771	0.6293	1.2349
7°	0.1219	0.9925	0.1228	52°	0.7880	0.6157	1.2799
8°	0.1392	0.9903	0.1405	53°	0.7986	0.6018	1.3270
9°	0.1564	0.9877	0.1584	54°	0.8090	0.5878	1.3764
10°	0.1736	0.9848	0.1763	55°	0.8192	0.5736	1.4281
11°	0.1908	0.9816	0.1944	56°	0.8290	0.5592	1.4826
12°	0.2079	0.9781	0.2126	57°	0.8387	0.5446	1.5399
13°	0.2250	0.9744	0.2309	58°	0.8480	0.5299	1.6003
14°	0.2419	0.9703	0.2493	59°	0.8572	0.5150	1.6643
15°	0.2588	0.9659	0.2679	60°	0.8660	0.5000	1.7321
16°	0.2756	0.9613	0.2867	61°	0.8746	0.4848	1.8040
17°	0.2924	0.9563	0.3057	62°	0.8829	0.4695	1.8807
18°	0.3090	0.9511	0.3249	63°	0.8910	0.4540	1.9626
19°	0.3256	0.9455	0.3443	64°	0.8988	0.4384	2.0503
20°	0.3420	0.9397	0.3640	65°	0.9063	0.4226	2.1445
21°	0.3584	0.9336	0.3839	66°	0.9135	0.4067	2.2460
22°	0.3746	0.9272	0.4040	67°	0.9205	0.3907	2.3559
23°	0.3907	0.9205	0.4245	68°	0.9272	0.3746	2.4751
24°	0.4067	0.9135	0.4452	69°	0.9336	0.3584	2.6051
25°	0.4226	0.9063	0.4663	70°	0.9397	0.3420	2.7475
26°	0.4384	0.8988	0.4877	71°	0.9455	0.3256	2.9042
27°	0.4540	0.8910	0.5095	72°	0.9511	0.3090	3.0777
28°	0.4695	0.8829	0.5317	73°	0.9563	0.2924	3.2709
29°	0.4848	0.8746	0.5543	74°	0.9613	0.2756	3.4874
30°	0.5000	0.8660	0.5774	75°	0.9659	0.2588	3.7321
31°	0.5150	0.8572	0.6009	76°	0.9703	0.2419	4.0108
32°	0.5299	0.8480	0.6249	77°	0.9744	0.2250	4.3315
33°	0.5446	0.8387	0.6494	78°	0.9781	0.2079	4.7046
34°	0.5592	0.8290	0.6745	79°	0.9816	0.1908	5.1446
35°	0.5736	0.8192	0.7002	80°	0.9848	0.1736	5.6713
36°	0.5878	0.8090	0.7265	81°	0.9877	0.1564	6.3138
37°	0.6018	0.7986	0.7536	82°	0.9903	0.1392	7.1154
38°	0.6157	0.7880	0.7813	83°	0.9925	0.1219	8.1443
39°	0.6293	0.7771	0.8098	84°	0.9945	0.1045	9.5144
40°	0.6428	0.7660	0.8391	85°	0.9962	0.0872	11.4301
41°	0.6561	0.7547	0.8693	86°	0.9976	0.0698	14.3007
42°	0.6691	0.7431	0.9004	87°	0.9986	0.0523	19.0811
43°	0.6820	0.7314	0.9325	88°	0.9994	0.0349	28.6363
44°	0.6947	0.7193	0.9657	89°	0.9998	0.0175	57.2900
45°	0.7071	0.7071	1.0000	90°	1.0000	0.0000	∞

Problem Solving Guide

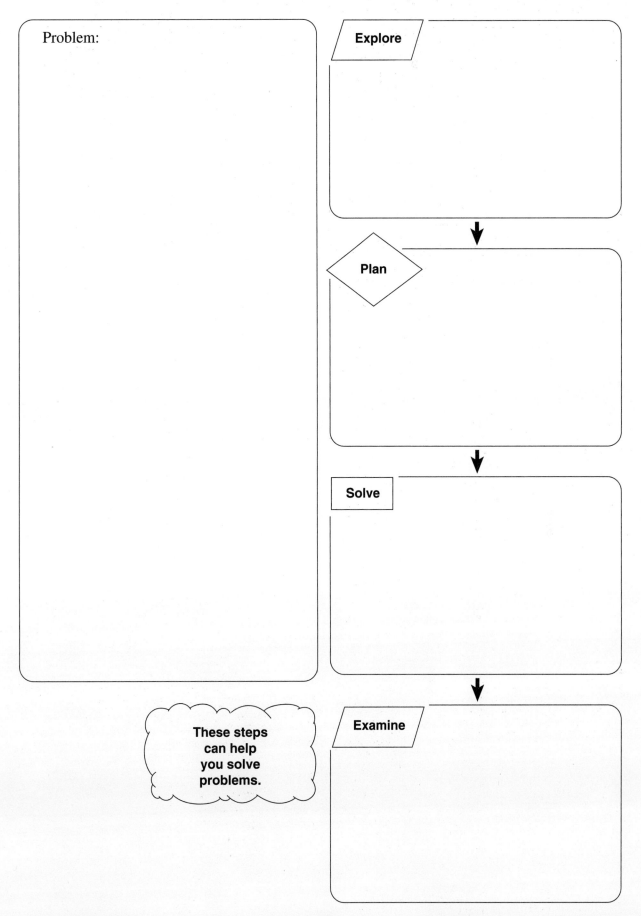

Chapter 1

Points, Lines, Planes, and Angles
Teaching Notes and Overview

Geometry Activity Recording Sheet
Modeling Intersecting Planes
(p. 26 of this booklet)

Use With the activity on page 8 in Lesson 1-1 of the Student Edition.

Objective Model the intersection of two planes and investigate points on the planes.

Materials
two index cards
scissors*
tape
* = available in Overhead Manipulative Resources

For this activity, students use index cards to illustrate the intersection of two planes. Be sure students label each plane and the line they have in common. Discuss the findings once students complete Exercises 1–4. You may wish to select students to share with the class where they drew each point on the planes.

Answers
See Teacher Wraparound Edition page 8.

Mini-Project
Intersecting Planes
(p. 27 of this booklet)

Use With Lesson 1-1.

Objective Model the intersection of three planes.

Materials
four sheets of heavy construction paper
scissors*
* = available in Overhead Manipulative Resources

Students should work in groups of two or three for this activity. They are to use two of the pieces of construction paper to complete steps 1–4 to illustrate the intersection of three planes. Then students answer questions from their illustration. For Exercise 5, students use the other two pieces of construction paper to create a different intersection of three planes. The students are guided to the conclusion that the intersection of three planes is a point.

Answers
1. $\overleftrightarrow{PR}$
2. $\overleftrightarrow{RS}$
3. $\overleftrightarrow{QR}$
4. point R; a point
5. See students' work.
6. Folded plane $\mathcal{A}$ sits straight with respect to plane $\mathcal{B}$, but folded plane $\mathcal{D}$ slants with respect to plane $\mathcal{C}$.

Geometry Activity Recording Sheet
Probability and Segment Measure
(p. 28 of this booklet)

Use With Lesson 1-2 as a follow-up activity. This corresponds to the activity on page 20 in the Student Edition.

Objective Find the probability that a random point lies in a segment contained by another segment.

Materials
none

This activity deals with finding probability involving a line segment. Remind students that

$$P(\text{event}) = \frac{\text{number of favorable outcomes}}{\text{total number of possible outcomes}}.$$

Go over the section titled Collect Data with the class. You may wish to have students work

© Glencoe/McGraw-Hill

Chapter 1 Teaching Notes and Overview

in pairs to complete Exercises 1–5. Remind students to state each probability as a reduced fraction.

Answers
See Teacher Wraparound Edition page 20.

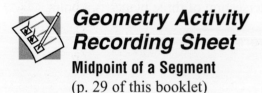

Geometry Activity Recording Sheet
Midpoint of a Segment
(p. 29 of this booklet)

Use With the activity on page 22 in Lesson 1-3 of the Student Edition.

Objective Use paper folding on a coordinate grid to locate the midpoint of a segment.

Materials
grid paper

For this activity, students can use the coordinate grid provided on the recording sheet, or a piece of grid paper. Students plot each pair of points and then fold the coordinate grid to find the midpoint of each segment. As students examine the coordinates of each midpoint, they should discover an algebraic method to finding the midpoint.

Answers
See Teacher Wraparound Edition page 22.

Using Overhead Manipulatives
Bisecting a Segment
(p. 30 of this booklet)

Use With Lesson 1-3.

Objective Bisect a line segment.

Materials
straightedge
transparency pens*
compass*
blank transparency
* = available in Overhead Manipulative Resources

This demonstration involves using a compass and a straightedge to construct a segment bisector on a blank transparency. You may wish to have students complete the construction at their desks while you complete the construction on the overhead. Ask students for suggestions of different ways you may be able to complete the construction.

Answers
Answers appear on the teacher demonstration instructions on page 30.

Geometry Activity
Pythagorean Puzzle
(pp. 31–33 of this booklet)

Use With Lesson 1-3.

Objective Use a puzzle to discover the Pythagorean Theorem.

Materials
classroom set of Geometry Activity worksheets
scissors*
blank transparency
* = available in Overhead Manipulative Resources

For this activity, students can work in pairs. Have students cut out the shapes on the worksheet. Instruct students to label the back of each shape with the title of the shape. Have students work with their partner to complete the activity. As you review the answers, select a student to form each of the four puzzles on the overhead. Discuss with students how their conjecture in Exercise 11 is a representation of the Pythagorean Theorem.

© Glencoe/McGraw-Hill   Teaching Geometry with Manipulatives

Chapter 1 Teaching Notes and Overview

Answers

1.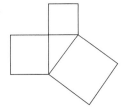

2. The length of a side of square A is equal to the length of one leg of the right triangle. The length of a side of square B is equal to the length of one leg of the right triangle. The length of a side of square C is equal to the length of the hypotenuse.

3.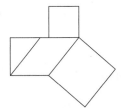

4. They are equal.

5.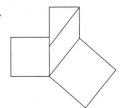

6. They are equal.

7.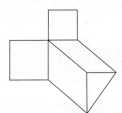

8. They are equal.
9. area of parallelogram B; area of parallelogram A; area of square C
10. area of square A + area of square B = area of square C
11. The area of the square on one leg plus the area of the square on the other leg is equal to the area of the square on the hypotenuse.

 Geometry Activity Recording Sheet
Modeling the Pythagorean Theorem
(p. 34 of this booklet)

Use With Lesson 1-3 as a follow-up activity. This corresponds to the activity on page 28 in the Student Edition.

Objective Use an area model to discover the Pythagorean Theorem.

Materials
grid paper
straightedge

In this activity, students draw right triangles and squares that attach to and represent the sides of the triangles. By examining the areas of the squares, which represent the square of each side, students discover the Pythagorean Theorem. This activity should serve as a brief overview of the Pythagorean Theorem, as it will be covered in greater depth in Chapter 7.

Answers
See Teacher Wraparound Edition page 28.

 Using Overhead Manipulatives
Measuring Angles
(p. 35 of this booklet)

Use With Lesson 1-4.

Objective Measure angles by using a protractor.

Materials
protractor*
transparency pens*
transparency prepared as described below
* = available in Overhead Manipulative Resources

© Glencoe/McGraw-Hill 23 Teaching Geometry with Manipulatives

Chapter 1 Teaching Notes and Overview

This demonstration involves using a protractor to measure an angle. You may wish to draw several angles of different measures on blank transparencies. Then ask students to come to the overhead and use a protractor to measure each angle.

In the extension, students are asked to use a protractor to draw an angle that measures 125°. You may wish to first demonstrate on the overhead how to use a protractor to draw an angle of a given measure other than 125°.

Answers
Answers appear on the teacher demonstration instructions on page 35.

Using Overhead Manipulatives
Constructing Congruent Angles
(p. 36 of this booklet)

Use With Lesson 1-4.

Objective Construct an angle congruent to a given angle.

Materials
transparency pens*
compass*
straightedge
protractor*
blank transparency
* = available in Overhead Manipulative Resources

This demonstration involves using a compass and straightedge to construct an angle congruent to a given angle. You may wish to review with students how to use a compass. Remind students that it is important to keep the same compass setting in order for the angles to be congruent. The protractor is used as a way to check your construction. Measure each angle with the protractor to verify that they are congruent.

Answers
Answers appear on the teacher demonstration instructions on page 36.

Geometry Activity Recording Sheet
Bisect an Angle
(p. 37 of this booklet)

Use With the activity on page 32 in Lesson 1-4 of the Student Edition.

Objective Use patty paper or tracing paper to bisect an angle.

Materials
patty paper or tracing paper
protractor*
* = available in Overhead Manipulative Resources

In this activity, students draw an angle on a piece of patty paper or tracing paper. Then students fold the angle so the two rays forming the angle are aligned. By using a protractor to measure the angles, students discover that the two angles created by the fold are congruent, and equal to half the measure of the original angle.

Answers
See Teacher Wraparound Edition page 32.

Using Overhead Manipulatives
Constructing Angle Bisectors
(p. 38 of this booklet)

Use With Lesson 1-4.

Objective Construct the bisector of a given angle.

Chapter 1 Teaching Notes and Overview

Materials
transparency pens*
compass*
straightedge
protractor*
blank transparency
* = available in Overhead Manipulative Resources

This demonstration involves using a compass and straightedge to bisect an angle. You may wish to have students complete the construction at their desks while you complete the construction on the overhead. Have students draw any size angle to bisect. They can perform the same steps that you do at the overhead. Have students use a protractor to measure each half of the bisected angle to make sure they performed the construction correctly.

This extension involves using a straightedge and compass to construct an angle that is twice the size of a given angle. After you complete the construction at the overhead, you may wish to have students complete the construction at their desks with any angle they draw.

Answers
Answers appear on the teacher demonstration instructions on page 38.

Geometry Activity Recording Sheet
Angle Relationships
(p. 39 of this booklet)

Use With the activity on page 38 in Lesson 1-5 of the Student Edition.

Objective Use patty paper or tracing paper to explore the relationships among vertical angles and linear pairs.

Materials
patty paper or tracing paper
protractor*
* = available in Overhead Manipulative Resources

In this activity, students can work in pairs. Students use patty paper or tracing paper to fold and draw intersecting lines. Students fold the patty paper or tracing paper to discover that vertical angles are congruent, and that a linear pair has a sum of 180. Once students have completed the activity, ask volunteers to share the "rules" they wrote in Exercise 6 about vertical angles and linear pairs.

Answers
See Teacher Wraparound Edition page 38.

Geometry Activity Recording Sheet
Constructing Perpendiculars
(p. 40 of this booklet)

Use With Lesson 1-5 as a follow-up activity. This corresponds to the activity on page 44 in the Student Edition.

Objective Construct a line perpendicular to a given line through a point on the line and through a point not on the line.

Materials
compass*
straightedge
* = available in Overhead Manipulative Resources

As a follow-up to Lesson 1-5, students construct perpendicular lines, both through a point on the line and a point not on the line. You may wish to demonstrate each construction at the overhead first. Then have students use the recording sheet to construct the perpendicular lines on their own. Once students have completed the activity, discuss how the two constructions are similar and different.

Answers
See Teacher Wraparound Edition page 44.

© Glencoe/McGraw-Hill   Teaching Geometry with Manipulatives

NAME _____ DATE _____ PERIOD ____

Geometry Activity Recording Sheet

(Use with the activity on page 8 in Lesson 1-1 of the Student Edition.)

Modeling Intersecting Planes

Materials
two index cards
scissors
tape

Analyze

1. Draw a point F on your model so that it lies in Q but not in R. Can F lie on $\overline{DC}$?

2. Draw point G so that it lies in R, but not in Q. Can G lie on $\overline{DC}$?

3. If point H lies in both Q and R, where would it lie? Draw point H on your model.

4. Draw a sketch of your model on paper. Label all points, lines, and planes appropriately.

© Glencoe/McGraw-Hill   Teaching Geometry with Manipulatives

NAME _____ DATE _____ PERIOD ____

Mini-Project

(Use with Lesson 1-1.)

Intersecting Planes

In order to visualize the intersection of three planes, make a model. Use four sheets of heavy construction paper and scissors.

Step 1: Label the four sheets plane A, plane B, plane C, and plane D.

Step 2: Fold plane A in half to form planes A_1 and A_2.

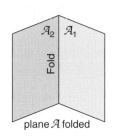

plane A folded

Step 3: On plane B, draw an angle whose segments are the same length as the edges of the angle formed by planes A_1 and A_2.

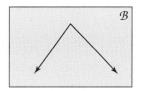

Step 4: Use your scissors to cut slits along the rays. Slip planes A_1 and A_2 through the slits. Label the angle in plane B as shown in the figure at the right.

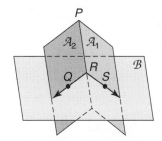

Answer each of the following.

1. What is the intersection of plane A_1 and plane A_2?
2. What is the intersection of plane A_1 and plane B?
3. What is the intersection of plane A_2 and plane B?
4. What do the answers to Exercises 1–3 have in common? What is the intersection of three planes?
5. Use planes C and D to construct a model in which the planes intersect as shown.
6. How does your construction differ from the first construction?

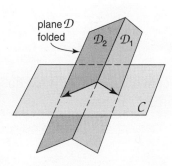

© Glencoe/McGraw-Hill   27   Teaching Geometry with Manipulatives

Geometry Activity Recording Sheet

(Use with the Lesson 1-2 Follow-Up Activity on page 20 in the Student Edition.)

Probability and Segment Measure

Materials
none

Analyze
Refer to the figure at the right.

1. Point J is contained in $\overline{WZ}$. What is the probability that J is contained in $\overline{XY}$?

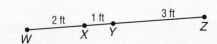

2. Point R is contained in $\overline{WZ}$. What is the probability that R is contained in $\overline{YZ}$?

3. Point S is contained in $\overline{WY}$. What is the probability that S is contained in $\overline{XY}$?

Make a Conjecture
Refer to the figure above.

4. Point T is contained in both $\overline{WY}$ and $\overline{XZ}$. What do you think is the probability that T is contained in $\overline{XY}$? Explain.

5. Point U is contained in $\overline{WX}$. What do you think is the probability that U is contained in $\overline{YZ}$? Explain.

NAME _____ DATE _____ PERIOD ____

Geometry Activity Recording Sheet

(Use with the activity on page 22 in Lesson 1-3 of the Student Edition.)

Midpoint of a Segment

Materials
grid paper

Make a Conjecture

1. What are the coordinates of point C?

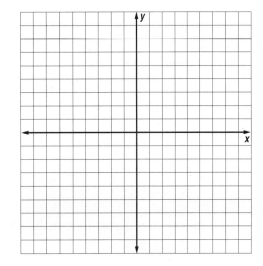

2. What are the lengths of $\overline{AC}$ and $\overline{CB}$?

3. What are the coordinates of point Z?

4. What are the lengths of $\overline{XZ}$ and $\overline{ZY}$?

5. Study the coordinates of points A, B, and C. Write a rule that relates these coordinates. Then use points X, Y, and Z to verify your conjecture.

Using Overhead Manipulatives
(Use with Lesson 1-3.)

Bisecting a Segment

Objective Bisect a line segment.

Materials
- straightedge
- transparency pens*
- compass*
- blank transparency

* = available in Overhead Manipulative Resources

Demonstration
Bisect a Segment
- Use the straightedge to draw a line segment. Label it $\overline{PQ}$.

- Open the compass to a setting that is longer than half the length of $\overline{PQ}$. Place the compass at P and draw a large arc.

- Using the same setting, place the compass at Q and draw a large arc to intersect the first.

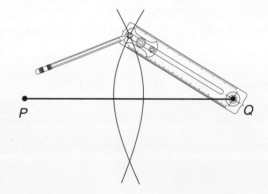

- Use the straightedge to draw a segment connecting the two intersection points. This segment intersects $\overline{PQ}$. Label the point of intersection M.

- Tell students that M bisects $\overline{PQ}$. Ask them how to verify this using a compass. **The same compass setting can be used for $\overline{PQ}$ and for $\overline{MQ}$.**

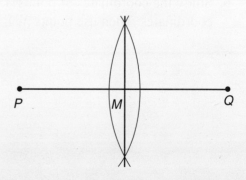

NAME _____ DATE _____ PERIOD ____

Geometry Activity
(Use with Lesson 1-3.)

Pythagorean Puzzle

Square A

Square B

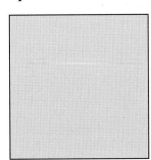

Square C

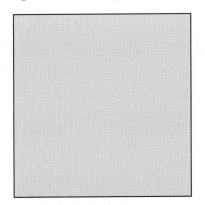

Triangle

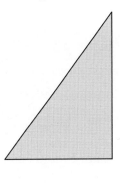

Parallelogram A

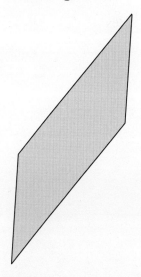

Parallelogram B

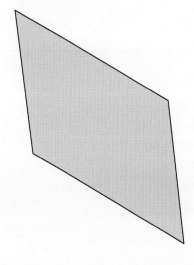

© Glencoe/McGraw-Hill 31 Teaching Geometry with Manipulatives

Geometry Activity

Cut out each figure on the previous page. Write the title for each shape on the back of the figure. Use the outline below to form each puzzle.

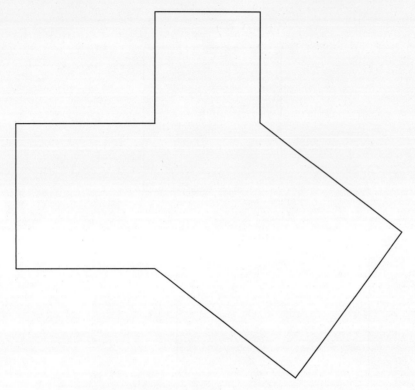

1. Use the triangle and square A, square B, and square C to form the puzzle. Make a drawing of the puzzle.

2. Compare the lengths of the sides of the squares with the lengths of the sides of the triangle.

3. Use the triangle, square A, square C, and parallelogram B to form the puzzle. Make a drawing of the puzzle.

Geometry Activity

4. What can you say about the area of parallelogram B and the area of square B?

5. Use the triangle, square B, square C, and parallelogram A to form the puzzle. Make a drawing of the puzzle.

6. What can you say about the area of parallelogram A and the area of square A?

7. Use the triangle, square A, square B, parallelogram A, and parallelogram B to form the puzzle. Make a drawing of the puzzle.

8. What can you conclude about the area of square C and the area of parallelogram A and parallelogram B?

9. Refer back to your conclusions in Exercises 4, 6, and 8 to complete each statement.

   area of square B = _____

   area of square A = _____

   area of parallelogram A + area of parallelogram B = _____

10. Substitute equal values of the last equation in Exercise 9 to write an equation that uses only the area of the squares.

11. Use the equation in Exercise 10 to make a conjecture about right triangles.

NAME _____ DATE _____ PERIOD ____

Geometry Activity Recording Sheet

(Use with the Lesson 1-3 Follow-Up Activity on page 28 in the Student Edition.)

Modeling the Pythagorean Theorem

Materials
grid paper
straightedge

Analyze

1. Determine the number of grid squares in each square you drew.

2. How do the numbers of grid squares relate?

3. If $AB = c$, $BC = a$, and $AC = b$, write an expression to describe each of the squares.

4. How does this expression compare with what you know about the Pythagorean Theorem?

Make a Conjecture

4. Repeat the activity for triangles with each of the side measures listed below. What do you find is true of the relationship of the squares on the sides of the triangle?

   a. 3, 4, 5   **b.** 8, 15, 17   **c.** 6, 8, 10

5. Repeat the activity with a right triangle whose shorter sides are both 5 units long. How could you determine the number of grid squares in the larger square?

© Glencoe/McGraw-Hill   Teaching Geometry with Manipulatives

Using Overhead Manipulatives
(Use with Lesson 1-4.)

Measuring Angles

Objective Measure angles by using a protractor.

Materials
- protractor*
- transparency pens*
- transparency prepared as described below

* = available in Overhead Manipulative Resources

Demonstration
Measure an Angle
- Prepare a transparency with the labeled angle as shown at the right. (Hint: Make the sides long enough to extend beyond the edge of the protractor. Answers will be given for an angle measuring 55°.)

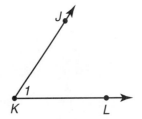

- Show students the transparency. Ask students to name the angle. **∠K, ∠JKL, ∠LKJ, or ∠1**
- Place the protractor over ∠JKL with the center point on vertex K. Discuss the markings on the protractor in relation to ∠JKL. Point out that the 0° line is not the same as the bottom of the protractor. Show students how you align the 0° mark on the protractor with side $\overrightarrow{KL}$.
- Locate the point on the protractor where $\overrightarrow{KJ}$ intersects the edge of the protractor. Ask students which edge to use when measuring this angle. **The scale that begins with 0 on the side where the 0° line is aligned with side $\overrightarrow{KL}$.** Have them read the measure of ∠JKL. **55°**

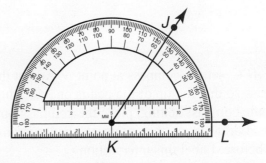

Extension
Draw an Angle
- Tell students they can also use a protractor to draw an angle with a specific measure. Have them draw ∠MNO if m∠MNO = 125°.

© Glencoe/McGraw-Hill   Teaching Geometry with Manipulatives

Using Overhead Manipulatives
(Use with Lesson 1-4.)

Constructing Congruent Angles

Objective Construct an angle congruent to a given angle.

Materials
- transparency pens*
- compass*
- straightedge
- protractor*
- blank transparency

*= available in Overhead Manipulative Resources

Demonstration
Construct Congruent Angles
- Use the straightedge to draw an angle. Label it ∠ABC. Beside ∠ABC, use the straightedge to draw $\overline{XY}$.

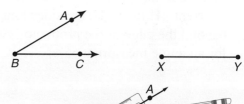

- Place the compass point at B. Draw an arc that intersects the sides of ∠ABC. Label the intersections M and N.

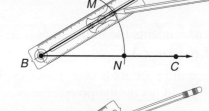

- Show students that you are keeping the same compass setting. Place the compass point at X and draw an arc that intersects $\overline{XY}$. Label this intersection P.

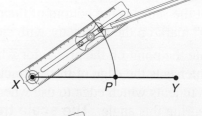

- On ∠ABC, set the compass at point N. Adjust the compass so that the tip of the pen is on M. Using the same compass setting, place the compass point at P, and draw an arc that intersects the larger arc you drew before. Label this intersection Q.
- Use the straightedge to draw $\overrightarrow{XQ}$.

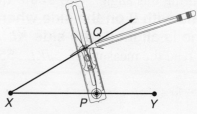

- Use the protractor to measure ∠ABC and ∠QXP. Ask students if the angles are congruent. **yes** Ask students to describe the procedure you used to copy ∠ABC on $\overline{XY}$. **Sample answer: Find two points the same distance from B and copy the distance on $\overline{XY}$. Find the distance from M to N and copy the distance on the arc. Find the point where the arcs intersect and draw the ray.**

© Glencoe/McGraw-Hill   36   Teaching Geometry with Manipulatives

NAME _____ DATE _____ PERIOD _____

Geometry Activity Recording Sheet

(Use with the activity on page 32 in Lesson 1-4 of the Student Edition.)

Bisect an Angle

Materials
patty paper or tracing paper
protractor

Analyze the Model

1. What seems to be true about ∠XYW and ∠WYZ?

2. Measure ∠XYZ, ∠XYW, and ∠WYZ.

3. You learned about a segment bisector in Lesson 1-3. Write a sentence to explain the term *angle bisector*.

Using Overhead Manipulatives
(Use with Lesson 1-4.)

Constructing Angle Bisectors

Objective Construct the bisector of a given angle.

Materials
- transparency pens*
- compass*
- straightedge
- protractor*
- blank transparency

* = available in Overhead Manipulative Resources

Demonstration
Construct the Bisector of an Angle

- Use the straightedge to draw any angle. Label the vertex B. Place the compass at the vertex and draw a large arc that intersects both sides of the angle. Label these points A and C.

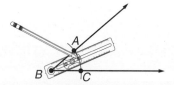

- Place the compass point at A and draw an arc to the right of A.
- Using the same compass setting and placing the compass point at C, draw an arc to intersect the one drawn from A. Label the intersection D. Ask students if the setting of the compass can be any length. **No; it must be longer than half the distance from A to C.**

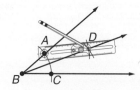

- Draw $\overrightarrow{BD}$. Tell students that $\overrightarrow{BD}$ bisects $\angle ABC$. Ask them how to verify this. **Measure $\angle ABD$ and $\angle DBC$ with a protractor.**

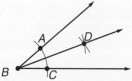

Extension
Double the Measure of an Angle

- Draw an acute angle and label the vertex Y. Draw a large arc through the sides of the angle and continue it to the left of Y. Label the intersection points X and Z.

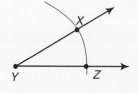

- Put the compass point at X. Adjust the setting so that it measures the distance from X to Z. Without removing the compass, draw an arc to intersect the large arc. Call this point U. Draw $\overrightarrow{YU}$. Ask students to describe the relationship between the measures of $\angle UYZ$ and $\angle XYZ$. **The measure of $\angle UYZ$ is twice the measure of $\angle XYZ$.**

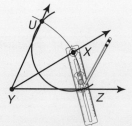

© Glencoe/McGraw-Hill   Teaching Geometry with Manipulatives

NAME _____ DATE _____ PERIOD ____

Geometry Activity Recording Sheet

(Use with the activity on page 38 in Lesson 1-5 of the Student Edition.)

Angle Relationships

Materials
patty paper or tracing paper
protractor

Analyze the Model

1. What do you notice about the angles when you made the last fold?

2. Fold the paper again through C so that $\overrightarrow{CB}$ aligns with $\overrightarrow{CE}$. What do you notice?

3. Use a protractor to measure each angle. Label the measure on your model.

4. Name pairs of vertical angles and their measures.

5. Name linear pairs of angles and their measures.

6. Compare your results with those of your classmates. Write a "rule" about the measures of vertical angles and another about the measures of linear pairs.

© Glencoe/McGraw-Hill   Teaching Geometry with Manipulatives

NAME _____ DATE _____ PERIOD ____

Geometry Activity Recording Sheet

(Use with the Lesson 1-5 Follow-Up Activity on page 44 in the Student Edition.)

Constructing Perpendiculars

Materials
compass
straightedge

Model and Analyze

1. Construct a line perpendicular to line *a* through a point on the line.

a ←――――――――――――――→

Construct a line perpendicular to line *b* through a point *not* on the line.

b ←――――――――――――――→

2. How is the second construction similar to the first one?

© Glencoe/McGraw-Hill   40   Teaching Geometry with Manipulatives

Chapter 2

Reasoning and Proof
Teaching Notes and Overview

Geometry Activity
If-Then Statements
(pp. 43–45 of this booklet)

Use With Lesson 2-3.

Objective Identify the if-then forms of conditional statements. State the converse of a conditional statement.

Materials
classroom set of Geometry Activity worksheets
scissors*
ruler*

*= available in Overhead Manipulative Resources

For this activity, divide the class into groups of three or four. Students first cut out the two triangles on the first page of the activity. Students then tear the angles from each triangle and form them in a straight line. This leads to the conclusion that the sum of the angles of a triangle is 180. While students complete the activity, encourage them to look for angle bisectors, how many right angles or obtuse angles are possible in one triangle, and the sum of the measures of two acute triangles. As an extension, you may wish to have students use the conditional statements discussed in this activity to write formal two-column proofs.

Answers
1. 180;

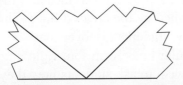

2. If a polygon is a triangle, then the sum of the measures of the angles is 180.

3. Sample answer: The sum of the 2 acute angles in a right triangle is 90°.

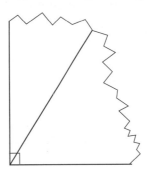

4. Sample answer: If a triangle is a right triangle, then the sum of the two acute angles is 90°.

5. Sample answer: If the sum of the two acute angles is 90°, then the triangle is a right triangle; true.

6. Sample answer: The supplement of the obtuse angle is the sum of the two acute angles.

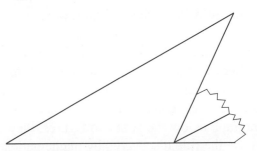

Geometry Activity Recording Sheet
Matrix Logic
(p. 46 of this booklet)

Use With Lesson 2-4 as a follow-up activity. This corresponds to the activity on page 88 in the Student Edition.

Objective Use matrix logic to organize information and solve problems.

Materials
none

Chapter 2 Teaching Notes and Overview

For this activity, students can work in groups of two or three. Review the example as a class before students work in groups to complete Exercises 1–2. The tables on the recording sheet will help students answer each exercise. You may wish to select two groups to explain how they arrived at their answers.

Answers
See Teacher Wraparound Edition page 88.

Mini-Project
Tracing Strategy
(p. 47 of this booklet)

Use With Lesson 2-4.

Objective Determine whether a figure can be traced without picking up your pencil and without tracing the same segment twice.

For this activity, students should work in groups of two or three. Students should examine the first two figures to determine whether they can be traced without picking up their pencil and without tracing the same segment twice. Once students read the paragraph below the first two figures, they can use the rule discovered by mathematicians to determine whether Exercises 1–3 can be traced in such a manner.

Answers
1. yes; X

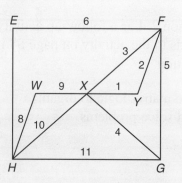

2. yes; A

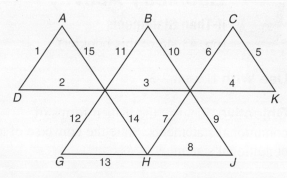

3. no

Geometry Activity Recording Sheet
Right Angles
(p. 48 of this booklet)

Use With the activity on page 110 in Lesson 2-8 of the Student Edition.

Objective Use paper folding to discover properties of perpendicular lines.

Materials
unlined paper
protractor*
* = available in Overhead Manipulative Resources

In this activity, students fold a piece of paper according to the directions given in the Student Edition. After opening the paper, students use a protractor to measure the angles formed by the creased lines. Through the activity, students discover that perpendicular lines form four right angles. You may wish to ask students how they know that the lines created by the paper folding are perpendicular lines.

Answers
See Teacher Wraparound Edition page 110.

© Glencoe/McGraw-Hill 42 Teaching Geometry with Manipulatives

NAME _____ DATE _____ PERIOD ___

Geometry Activity

(Use with Lesson 2-3.)

If-Then Statements

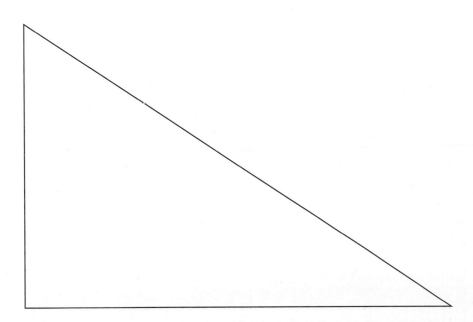

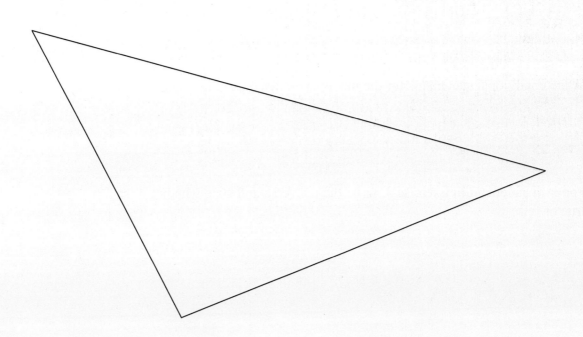

Geometry Activity

Cut out the triangles on the previous page. Tear the angles from each triangle as shown at the right.

1. What is the sum of the measures of the angles of the triangle?

   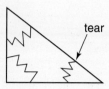

   At the right, draw a picture of the model that leads to your conclusion.

2. Write your answer to Exercise 1 as a conditional statement in the if-then form.

3. Each of your cutout triangles has two acute angles. Use the models to draw a conclusion about the relationship of the angles. Illustrate your models below.

4. Write conditional statements for the models you discovered in Exercise 3.

5. For each statement in Exercise 4, write the converse and state whether it is true or false.

Geometry Activity

6. Cut out one each of triangles A and B from the bottom of this page. Tear off the angles as you did for Exercises 1–5. As a group, make a hypothesis regarding each triangle. Then make a model of the hypothesis and draw a conclusion. Record the conditional statements in if-then form and explain why your statement is true. **NOTE:** You may compare the measure of each angle of the triangle with its supplement by extending a side of the triangle.

Triangle A   **Triangle B**

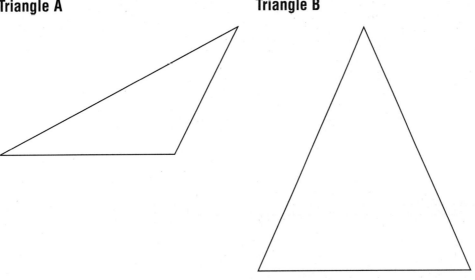

Triangle A   **Triangle B**

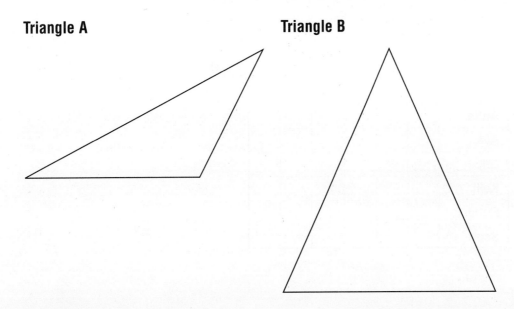

NAME _____ DATE _____ PERIOD ____

Geometry Activity Recording Sheet

(Use with the Lesson 2-4 Follow-Up Activity on page 88 in the Student Edition.)

Matrix Logic

Materials
none

Exercises
Use matrix logic and each table to answer Exercises 1–2.

1. Nate, John, and Nick just began after-school jobs. One works at a veterinarian's office, one at a computer store, and one at a restaurant. Nate buys computer games on the way to work. Nick is allergic to cat hair. John receives free meals at his job. Who works at which job?

After-School Job	Nate	John	Nick
Veterinarian's office			
Computer store			
Restaurant			

2. Six friends live in consecutive apartments on the same side of their apartment building. Anita lives in apartment C. Kelli's apartment is just past Scott's. Anita's closest neighbors are Eric and Ava. Scott's apartment is not A through D. Eric's apartment is before Ava's. If Roberto lives in one of the apartments, who lives in which apartment?

Friend	A	B	C	D	E	F
Anita						
Ava						
Eric						
Kelli						
Roberto						
Scott						

NAME _____ DATE _____ PERIOD ____

Mini-Project

(Use with Lesson 2-4.)

Tracing Strategy

Try to trace over each of the figures below without picking up your pencil and without tracing the same segment twice.

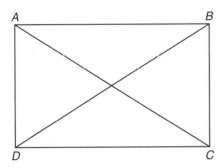 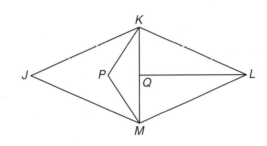

The figure at the left cannot be traced, but the one at the right can. Mathematicians have proved that a figure is traceable if it has no more than two points where an odd number of segments meet. The figure at the left has three segments meeting at each of the four corners. However, the figure at the right has only two points, L and Q, where an odd number of segments meet.

Determine whether each figure can be traced. If it can, then name the starting point and number the sides in the order they should be traced.

1. 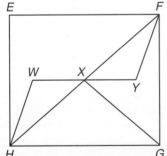   2.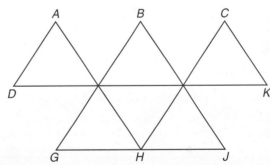

3.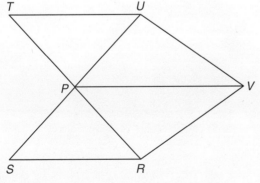

© Glencoe/McGraw-Hill   47   Teaching Geometry with Manipulatives

NAME _____ DATE _____ PERIOD _____

Geometry Activity Recording Sheet

(Use with the activity on page 110 in Lesson 2-8 of the Student Edition.)

Right Angles

Materials
unlined paper
protractor

Analyze the Model

1. What do you notice about the lines formed?

2. What do you notice about each pair of adjacent angles?

3. What are the measures of the angles formed?

Make a Conjecture

4. What is true about perpendicular lines?

5. What is true about all right angles?

Chapter 3 Parallel and Perpendicular Lines
Teaching Notes and Overview

Geometry Activity Recording Sheet
Draw a Rectangular Prism
(p. 52 of this booklet)

Use With the activity on page 126 in Lesson 3-1 of the Student Edition.

Objective Draw a rectangular prism and investigate the segments and planes that make up the prism.

Materials
none

For this activity, students draw a rectangular prism and then identify parallel planes, intersecting planes, and parallel segments of the figure. Students can use the blank space available on the recording sheet to draw the rectangular prism. You may wish to select a student to draw a prism on the chalkboard or overhead before you go over the answers to Exercises 1–3.

Answers
See Teacher Wraparound Edition page 126.

Using Overhead Manipulatives
Parallels and Transversals
(p. 53 of this booklet)

Use With Lesson 3-2.

Objective Identify the relationships between pairs of angles formed by pairs of parallel lines and transversals.

Materials
lined paper transparency*
transparency pens*
protractor*
straightedge

* = available in Overhead Manipulative Resources

This demonstration involves discovering the relationships among the angles created by two parallel lines cut by a transversal. Once all angle pairs are identified, draw a set of vertical parallel lines cut by a transversal. Number the angles differently and question the students about all the angle pairs formed.

Answers
Answers appear on the teacher demonstration instructions on page 53.

Mini-Project
Avoiding Lattice Points
(p. 54 of this booklet)

Use With Lesson 3-4.

Objective Make a conjecture about what determines whether the graph of an equation will pass through lattice points.

By examining the two equations and graphs, students should be able to write an equation with integer values for a, b, and c, that does not contain lattice points when graphed. Students should be guided to the conclusion that any equation for which the greatest common divisor of a and b does not divide c will result in an equation whose graph does not contain lattice points.

Answers
1. Infinitely many; explanations will vary. See students' work.

2. No; if n passed through a lattice point, you could put the coordinates of that point in the equation for line n and use the equation to express $\sqrt{2}$ as a rational number. This contradicts the fact that $\sqrt{2}$ is irrational.

3. Answers will vary. Any equation for which the greatest common divisor of a and b does not divide c will work.

© Glencoe/McGraw-Hill — Teaching Geometry with Manipulatives

Chapter 3 Teaching Notes and Overview

Geometry Activity
Graphing Lines in the Coordinate Plane
(pp. 55–56 of this booklet)

Use With Lesson 3-4.

Objective Graph parallel and perpendicular lines. Graph a line given its slope and y-intercept.

Materials
classroom set of Geometry Activity worksheets
grid paper transparency
Pick-Up Sticks®
8–10 heavyweight pieces of paper, spinners, and brass fasteners
8–10 number cubes*
glue or tape
* = available in Overhead Manipulative Resources

For this activity, each group should attach one spinner grid from the worksheet to the heavyweight paper, using tape or glue. Have students complete the spinner according to the directions on the worksheet. Introduce the game Y ∪ X, and have the groups play the game. You may wish to have students play the game according to one of the variations on the worksheet.

Using Overhead Manipulatives
Constructing Parallel Lines
(p. 57 of this booklet)

Use With Lesson 3-5.

Objective Construct a line parallel to a given line through a point not on the line.

Materials
straightedge
transparency pens*
compass*
protractor*
blank transparency
* = available in Overhead Manipulative Resources

The demonstration involves constructing parallel lines. You may wish to have a student come to the overhead and use a protractor to verify that the lines are parallel.

In this extension, construct another line parallel to the ones constructed. You may wish to have students complete a construction of parallel lines at their desk.

Answers
Answers appear on the teacher demonstration instructions on page 57.

Using Overhead Manipulatives
Constructing Perpendicular Lines
(pp. 58–59 of this booklet)

Use With Lesson 3-6.

Objective Construct a line perpendicular to another line through a point on the line or through a point not on the line.

Materials
transparency pens*
compass*
straightedge
blank transparency
* = available in Overhead Manipulative Resources

There are two demonstrations for this activity.
- Demonstration 1 involves constructing a line perpendicular to another line through a point on the line. Ask why it is necessary to

© Glencoe/McGraw-Hill   Teaching Geometry with Manipulatives

Chapter 3 Teaching Notes and Overview

open the compass wider before you perform Step 3.
- Demonstration 2 involves constructing a line perpendicular to another line through a point not on the line.
- The extension is a discussion of how perpendiculars can be used to construct a square. You may wish to actually construct a square on a blank transparency.

Answers
Answers appear on the teacher demonstration instructions on pages 58–59.

Using Overhead Manipulatives
Parallels and Distance
(pp. 60–62 of this booklet)

Use With Lesson 3-6.

Objective Find the distance between two parallel lines.

Materials
TI-83 Plus graphing calculator
TI ViewScreen™, if available, or a blank transparency prepared as described
transparency pens*
compass*
straightedge
coordinate grids transparency*
* = available in Overhead Manipulative Resources

This activity includes two demonstrations for finding the distance between parallel lines. If you have a TI ViewScreen™, complete Demonstration 1. If you do not, complete Demonstration 2 using the coordinate grids transparency. You may choose to complete both demonstrations to illustrate each method to students.
- Demonstration 1 involves using a graphing calculator to find the distance between two lines. Discuss the reason for graphing the perpendicular line when you are trying to find the distance between the parallel lines.
- Demonstration 2 involves graphing the two parallel lines from Demonstration 1, $y = 3.5x + 8$ and $y = 3.5x - 7$, on the coordinate grids transparency. Then by construction, you graph the perpendicular line.

Answers
Answers appear on the teacher demonstration instructions on pages 60–62.

Geometry Activity Recording Sheet
Non-Euclidean Geometry
(p. 63 of this booklet)

Use With Lesson 3-6 as a follow-up activity. This corresponds to the activity on pages 165–166 in the Student Edition.

Objective Investigate the relationship between spherical geometry and plane Euclidean geometry.

Materials
none

To begin this activity, discuss the similarities and differences between spherical geometry and plane Euclidean geometry on pages 165–166. Discuss the example with the class before having students divide into small groups to complete the exercises. You may wish to have a sphere with rubber bands available for each group to model a plane and lines in spherical geometry. While going over the answers to the exercises, you may wish to select students to demonstrate each property using the sphere and rubber bands.

Answers
See Teacher Wraparound Edition page 166.

NAME _____ DATE _____ PERIOD ____

Geometry Activity Recording Sheet

(Use with the activity on page 126 in Lesson 3-1 of the Student Edition.)

Draw a Rectangular Prism

Materials
none

Use the space provided to draw a rectangular prism.

Analyze

1. Identify the parallel planes in the figure.

2. Name the planes that intersect plane *ABC* and name their intersections.

3. Identify all segments parallel to $\overline{BF}$.

© Glencoe/McGraw-Hill — Teaching Geometry with Manipulatives

Using Overhead Manipulatives
(Use with Lesson 3-2.)

Parallels and Transversals

Objective Identify the relationships between pairs of angles formed by pairs of parallel lines and transversals.

Materials
- lined paper transparency*
- transparency pens*
- protractor*
- straightedge

* = available in Overhead Manipulative Resources

Demonstration
Identify Relationship among Parallels and Transversals
- Show students the lined paper transparency. Tell them that the lines on the paper are all parallel to each other. Use the straightedge to draw two parallel lines using the lines on the transparency. Then draw a line, like the one shown below, that intersects the two parallel lines. (*Hint:* Make the parallel lines far enough apart to easily measure the angles formed by the transversal. Answers will be given for a transversal forming 55° and 125° angles.)
- Label the angles formed using the numbers 1 through 8. Have students measure each angle and record its measurement at the side of the transparency.
  $m\angle 1 = 55$, $m\angle 2 = 125$, $m\angle 3 = 55$, $m\angle 4 = 125$,
  $m\angle 5 = 55$, $m\angle 6 = 125$, $m\angle 7 = 55$, $m\angle 8 = 125$

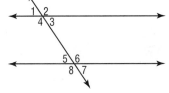

- On the drawing, use a different colored transparency pen to circle the numbers of all the angles that are congruent. **∠1, ∠3, ∠5, and ∠7 are congruent; ∠2, ∠4, ∠6, and ∠8 are congruent.**
- Tell students that ∠3, ∠4, ∠5, and ∠6 are examples of **interior angles**. Ask students which pairs of interior angles are congruent. **∠3 and ∠5 are congruent; ∠4 and ∠6 are congruent.** Tell them that congruent pairs of interior angles are called **alternate interior angles**.
- Ask students, "Which pairs of interior angles are supplementary?" **∠3 and ∠6 are supplementary; ∠4 and ∠5 are supplementary.** Tell them that interior supplementary angles are called **consecutive interior angles**.
- Tell students that ∠1, ∠2, ∠7, and ∠8 are examples of exterior angles. Ask students which pairs of exterior angles are congruent. **∠1 and ∠7 are congruent; ∠2 and ∠8 are congruent.** Tell them that congruent pairs of exterior angles are called **alternate exterior angles.**
- Ask students, "Which pairs of congruent angles are on the same side of the transversal?" **∠1, ∠5; ∠4, ∠8; ∠2, ∠6; ∠3, ∠7** Tell students that pairs of congruent angles that are on the same side of the transversal are called **corresponding angles**.

© Glencoe/McGraw-Hill

NAME _____ DATE _____ PERIOD ____

Mini-Project
(Use with Lesson 3-4.)

Avoiding Lattice Points

Points that have integers for both coordinates are called **lattice points**. Study the graph below.

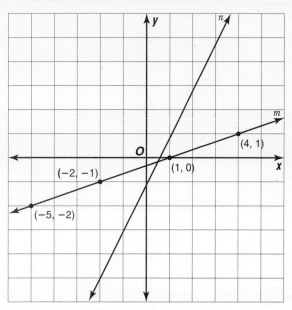

Line *m* has equation $x - 3y = 1$, and line *n* has equation $(30\sqrt{2})x - 21y = 28$.

Use the equations for lines *m* and *n* to help answer each exercise.

1. How many lattice points in the entire coordinate plane does line *m* pass through? Explain your answer.

2. Does line *n* ever pass through a lattice point? Explain your answer. (*Hint*: $\sqrt{2}$ is irrational.)

3. Write equations of the form $ax + by = c$, where *a*, *b*, and *c* are integers, that have graphs that never pass through lattice points.

© Glencoe/McGraw-Hill   Teaching Geometry with Manipulatives

NAME _____ DATE _____ PERIOD ____

Geometry Activity

(Use with Lesson 3-4.)

Graphing Lines in the Coordinate Plane

Y ∪ X Graph Game

A game of graphing lines on the Cartesian Coordinate System

Players: 2 to 4

Getting Ready to Play:

1. Before beginning the game, each group must construct a spinner. See the instructions on the next page.

2. Each player chooses 4 or 5 Pick-Up Sticks® to use for graphing his/her lines. Each player should choose a different color, when possible.

3. Each player spins the Y ∪ X spinner. The player who spins a slope closer to zero plays first.

Rules:

1. Player #1 will spin the Y ∪ X spinner and determine the slope of the spin arrow. Next, the player will roll the number cube for a *y*-intercept.

   Example:

   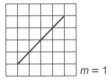 $m = 1$ $b = 2$

2. The player then graphs the line by placing a Pick-Up Stick® on the game board.

3. Players #2, 3, and 4 each take their turns.

4. The winner is determined by the first player to graph a line perpendicular or parallel to an axis or another line already graphed on the board.

Variations:

1. Players may wish to record the linear equations to verify the winning line algebraically.

2. Points may be awarded for parallel and perpendicular lines until all the sticks have been placed on the board. The player with the most points wins.

Geometry Activity

Spinner

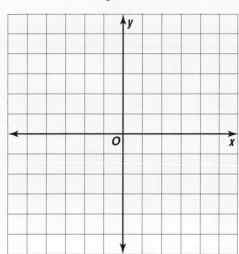

Assembly of Spinner

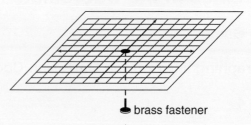

brass fastener

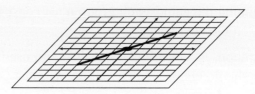

Game Board

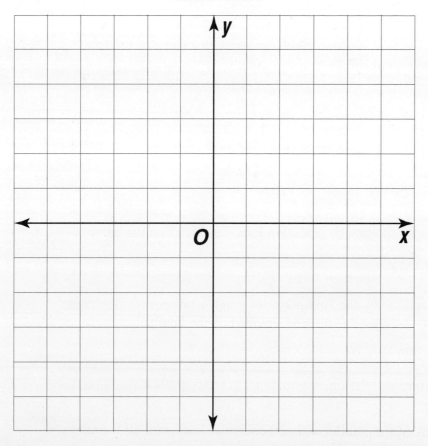

Using Overhead Manipulatives
(Use with Lesson 3-5.)

Constructing Parallel Lines

Objective Construct a line parallel to a given line through a point not on the line.

Materials
- straightedge
- transparency pens*
- compass*
- protractor*
- blank transparency

* = available in Overhead Manipulative Resources

Demonstration
Construct a Parallel Line through a Point not on the Line
- Near the center of a blank transparency, draw a horizontal line 5 inches long. Label the line *m*.
- Use a different colored transparency pen for the next five steps. Mark a point *N*, not on *m*. Draw a line through *N* that intersects *m* as shown. Label the intersection point *D*.
- Place the compass point at *D* and draw an arc. Label the points *E* and *F* as shown below.
- With the same setting, place the compass point at *N* and draw an arc. Label point *G* as shown.

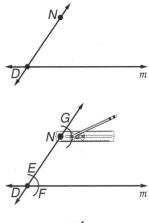

- Show students how you use the compass to measure the distance from *E* to *F*.
- With the same setting, place the compass point at *G* and draw an arc to intersect the one already drawn. Label this point *H*. Ask students how the distance from *E* to *F* compares to the distance from *G* to *H*. **It is the same.**
- Draw a line through *N* and *H* and label it line *n*. Say, "By construction, *n* is parallel to *m*."
- Review the types of congruent angles formed by parallel lines and a transversal. **alternate interior, alternate exterior, and corresponding** Ask students what type of angles you used to create the parallel lines. **corresponding angles**

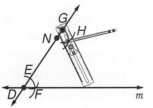

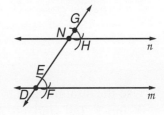

Extension
Construct Another Parallel Line
- Construct a line parallel to *m* below *n*. Label it *b*. Ask students what the relationship is between *n* and *b*. **They are parallel.**

© Glencoe/McGraw-Hill   57   Teaching Geometry with Manipulatives

Using Overhead Manipulatives
(Use with Lesson 3-6.)

Constructing Perpendicular Lines

Objective Construct a line perpendicular to another line through a point on the line or through a point not on the line.

Materials
- transparency pens*
- compass*
- straightedge
- blank transparency

* = available in Overhead Manipulative Resources

Demonstration 1
Construct a Line Perpendicular to Another Line through a Point on the Line

- Draw a line and label it *a*. Choose a point on the line and label it *X*.

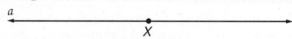

- Place the compass point on *X* and draw arcs to intersect line *a* on both sides of *X*. Label these points *R* and *S*.

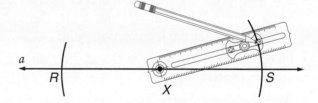

- Open the compass wider. Place the compass point at *R* and draw an arc above *X*. With the compass at the same setting, place the compass point at *S* and draw an arc that intersects the previous arc. Label this intersection point *Y*.

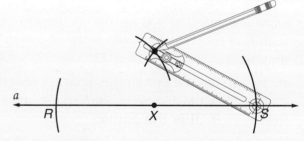

- Use the straightedge to draw a line through *X* and *Y*. Tell students that $\overleftrightarrow{XY}$ is perpendicular to line *a* by construction.

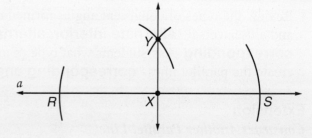

© Glencoe/McGraw-Hill

Using Overhead Manipulatives

Demonstration 2
Construct a Line Perpendicular to Another Line through a Point not on the Line

- Draw a line and label it *m*. Choose any point above *m* and label it *P*.

- Open the compass to a width greater than the shortest distance from *P* to *m*. Draw a large arc to intersect *m* twice. Label these points of intersection *T* and *U*.

- Place the compass point at *T* and draw an arc below *m*.
- Using the same compass setting, place the compass point at *U*. Draw an arc to intersect the one drawn from *T*. Label this point of intersection *A*.

- Use the straightedge to draw a line through *P* and *A*. Tell students that $\overleftrightarrow{PA}$ is perpendicular to line *m* by construction.

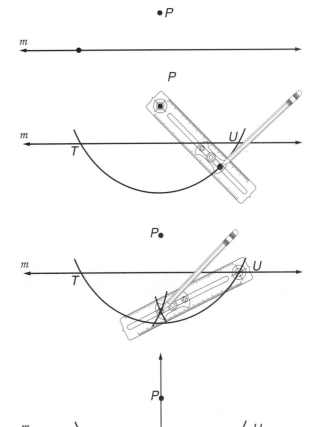

Extension
Use Perpendiculars to Construct a Square

- Ask students to describe how they could use perpendiculars to construct a square. **Sample answer: Choose any two points on a line. Label them *R* and *S*. Construct a perpendicular through *R*. Use a compass to measure the distance from *R* to *S*. Place the compass point at *R* and draw an arc on the perpendicular through *R*. Label the point *T*. Using the same setting, place the compass point at *T* and draw an arc above *S*. Then place the compass point at *S* and draw an arc to intersect the one drawn from *T*. Call this point *U*. Use a straightedge to draw $\overline{TU}$ and $\overline{US}$. Figure *RSUT* is a square.**

© Glencoe/McGraw-Hill 59 *Teaching Geometry with Manipulatives*

Using Overhead Manipulatives
(Use with Lesson 3-6.)

Parallels and Distance

Objective Find the distance between two parallel lines.

Materials
- TI-83 Plus graphing calculator
- TI ViewScreen™, if available, or a blank transparency prepared as described below
- transparency pens*
- compass*
- straightedge
- coordinate grids transparency*

* = available in Overhead Manipulative Resources

Demonstration 1
Find Distance Using a Graphing Calculator

- If a TI ViewScreen™ is not available, use Demonstration 2. If you have graphing calculators that are not TI-83s, adjust the given keystrokes for your calculator.

- Set the axes ranges by entering the following keystrokes (TI-83 Plus).
  [WINDOW] [(-)] 4 [ENTER] 5 [ENTER] 1 [ENTER] [(-)]
  3 [ENTER] 3 [ENTER] 1 [ENTER] 1 [2nd] [QUIT]

- Then graph $y = 3.5x + 8$, $y = 3.5x - 7$, and $y = -\frac{2}{7}x + \frac{4}{7}$ by entering the following keystrokes.
  [Y=] 3.5 [X,T,θ] [+] 8 [ENTER] 3.5 [X,T,θ] [-] 7 [ENTER]
  [(] [(-)] 2 [÷] 7 [)] [X,T,θ] [+] [(] 4 [÷] 7 [)]

- Tell students that the first two equations will graph parallel lines because their slopes are the same. Ask students to explain how the slope of the third line relates to the slope of the first two. **It is the negative reciprocal because** $3.5 = \frac{7}{2}$ **and the negative reciprocal of** $\frac{7}{2}$ **is** $-\frac{2}{7}$. Ask students how this will affect the graph of the third line. **It will be perpendicular to the other two lines.**

- Remind students that when a line intersects a pair of parallel lines it is called a transversal.

© Glencoe/McGraw-Hill   Teaching Geometry with Manipulatives

Using Overhead Manipulatives

- Press GRAPH to display the three lines.
- Press TRACE, then ▲ on the TI-83 Plus. Move the cursor left to the point where the transversal intersects the parallel line. Ask students what the coordinates of this point are. **(−2, 1)** Record the coordinates. Move the cursor right to the point where the transversal intersects the other parallel line. Ask students what the coordinates of this point are. **(2, 0)** Record the coordinates.

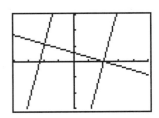

- Ask students to use the Distance Formula to find the distance between the two parallel lines. You may wish to refer students to the Distance Formula in Lesson 1-3 of *Glencoe Geometry*. **The distance between the parallel lines is $\sqrt{17}$ or about 4.1 units.**

Demonstration 2
Find Distance Using a Compass and Straightedge
- Copy the graphs shown on the coordinate grid transparency.

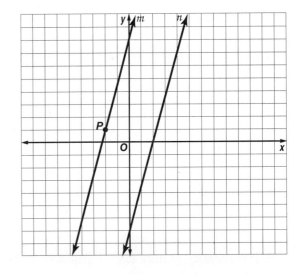

- Tell students that you are going to construct a line perpendicular to line *n* through point $P(-2, 1)$.
- Place the compass point at *P*. Make the setting wide enough so that when an arc is drawn, it intersects *n* in two places. Label these points of intersection *Q* and *R*.

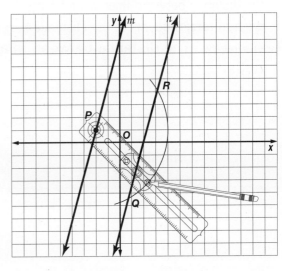

© Glencoe/McGraw-Hill 61 Teaching Geometry with Manipulatives

Using Overhead Manipulatives

- Using the same compass setting, place the compass point at R and draw an arc below the x-axis and to the right of the previous arc. Then, still using the same compass setting, place the compass point at Q and draw an arc to intersect the one drawn from R. Label the point of intersection S.

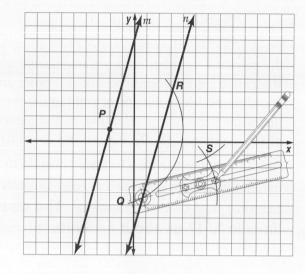

- Draw $\overleftrightarrow{PS}$. Tell students that $\overleftrightarrow{PS}$ is perpendicular to line m by construction.

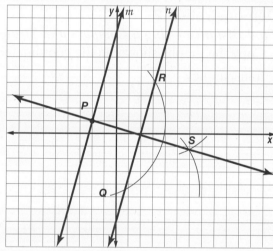

- Ask students, "Where does $\overleftrightarrow{PS}$ intersect line n?" **(2, 0)** Tell students to find the distance between lines m and n by using the Distance Formula. You may wish to refer students to the Distance Formula in Lesson 1-3 of *Glencoe Geometry*. **The distance between m and n is $\sqrt{17}$ or about 4.1 units.**

NAME _____ DATE _____ PERIOD ____

Geometry Activity Recording Sheet

(Use with the Lesson 3-6 Follow-Up Activity on pages 165–166 in the Student Edition.)

Non-Euclidean Geometry

Materials
none

Exercises
For each property from plane Euclidean geometry, write a corresponding statement for spherical geometry.

1. A line goes on infinitely in two directions.

2. A line segment is the shortest path between two points.

3. Two distinct lines with no point of intersection are parallel.

4. Two distinct intersecting lines intersect in exactly one point.

5. A pair of perpendicular straight lines divides the plane into four infinite regions.

6. Parallel lines have infinitely many common perpendicular lines.

7. There is only one distance that can be measured between two points.

If spherical points are restricted to be nonpolar points, determine if each statement from plane Euclidean geometry is also *true* in spherical geometry. If *false*, explain your reasoning.

8. Any two distinct points determine exactly one line.

9. If three points are collinear, exactly one point is between the other two.

10. Given a line ℓ and point P not on ℓ, there exists exactly one line parallel to ℓ passing through P.

© Glencoe/McGraw-Hill   Teaching Geometry with Manipulatives

Chapter 4
Congruent Triangles
Teaching Notes and Overview

Geometry Activity Recording Sheet
Equilateral Triangles
(p. 69 of this booklet)

Use With the activity on page 179 in Lesson 4-1 of the Student Edition.

Objective Use patty paper to form an equilateral triangle, isosceles triangle, and scalene triangle.

Materials
patty paper or tracing paper

For this activity, students arrange and fold patty paper to form an equilateral triangle, isosceles triangle, and scalene triangle. You may wish to select three students to demonstrate how they formed each triangle. Discuss with students how they can be sure that each triangle is equilateral, isosceles, or scalene.

Answers
See Teacher Wraparound Edition page 179.

Geometry Activity Recording Sheet
Angles of Triangles
(p. 70 of this booklet)

Use With Lesson 4-2 as a preview activity. This corresponds to the activity on page 184 in the Student Edition.

Objective Draw and manipulate triangles to discover the Angle Sum Theorem and Exterior Angle Sum Theorem.

Materials
unlined paper
straightedge
protractor*

* = available in Overhead Manipulative Resources

This activity has two parts. Students can work in pairs to complete each activity. In Activity 1, students draw and manipulate an obtuse triangle. By measuring angles, they can make a conjecture that the sum of the angle measures of any triangle is 180. In Activity 2, students trace the same triangle from Activity 1. By tearing off angles and matching them to exterior angles, they can make a conjecture that the measure of an exterior angle is equal to the sum of the measures of the two remote interior angles.

Answers
See Teacher Wraparound Edition page 184.

Using Overhead Manipulatives
Angle Measures in Triangles
(p. 71 of this booklet)

Use With Lesson 4-2.

Objective Discover the sum of the measures of the angles in a triangle.

Materials
transparency pens*
scissors*
tape
straightedge
paper
blank transparency

* = available in Overhead Manipulative Resources

This demonstration involves discovering that the sum of the angle measures of a triangle is 180. If you choose to have students complete the activity at their desks, you may wish to have a few students come to the overhead to demonstrate the activity with the triangle they created. After completing the demonstration, you may ask students how this demonstration is similar to a proof.

© Glencoe/McGraw-Hill 65 Teaching Geometry with Manipulatives

Chapter 4 Teaching Notes and Overview

Answers
Answers appear on the teacher demonstration instructions on page 71.

Using Overhead Manipulatives
Congruent Triangles
(p. 72 of this booklet)

Use With Lesson 4-3.

Objective Identify congruent triangles and corresponding parts of congruent triangles.

Materials
scissors*
transparency pens*
blank transparency
* = available in Overhead Manipulative Resources

This demonstration serves as an introduction for identifying and naming corresponding parts of congruent triangles. You may wish to create another example by drawing different triangles on the transparency and labeling them with different letters.

Answers
Answers appear on the teacher demonstration instructions on page 72.

Geometry Activity
Congruent Triangles
(pp. 73–74 of this booklet)

Use With Lesson 4-3.

Objective Identify the corresponding parts of congruent triangles.

Materials
classroom set of Geometry Activity worksheets
protractor*
tracing paper
scissors*
* = available in Overhead Manipulative Resources

Students begin this activity by measuring the angles of the three triangles on the worksheet. Then they prove that $\triangle ABC \cong \triangle XYZ$ and name their congruent parts. Students then trace and cut out the two smaller triangles. By arranging the triangles inside $\triangle QRS$, students discover different congruence statements and congruent parts. You may wish to have a few volunteers draw their illustrations from Exercises 3–5 on the chalkboard or overhead.

Answers
1. $m\angle A = 30$, $m\angle B = 60$, $m\angle C = 90$,
   $m\angle X = 30$, $m\angle Y = 60$, $m\angle Z = 90$,
   $AB = 4\frac{5}{16}"$, $BC = 2\frac{3}{16}"$, $AC = 3\frac{13}{16}"$,
   $XY = 4\frac{5}{16}"$, $YZ = 2\frac{3}{16}"$, $XZ = 3\frac{13}{16}"$

2. $\triangle ABC \cong \triangle XYZ$; $\angle A \cong \angle X$, $\angle B \cong \angle Y$, $\angle C \cong \angle Z$, $\overline{AB} \cong \overline{XY}$, $\overline{AC} \cong \overline{XZ}$, $\overline{BC} \cong \overline{YZ}$

3–5. Answers will vary. Drawings may be any of the following, or a variation of the following.

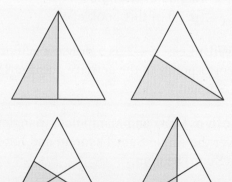

Chapter 4 Teaching Notes and Overview

Using Overhead Manipulatives
Tests for Congruent Triangles
(pp. 75–76 of this booklet)

Use With Lesson 4-4.

Objective Use SSS and SAS postulates to test for triangle congruence.

Materials
transparency pens*
compass*
straightedge
protractor*
5 blank transparencies
* = available in Overhead Manipulative Resources

This activity involves two demonstrations for creating congruent triangles.

- Demonstration 1 uses the SSS Postulate to prove that three constructed triangles are congruent. After you complete the first construction, you may wish to have students complete the next two constructions at their desks. You can then have a few students cut out their constructed triangles and lay them on the first triangle created on the overhead to verify that they are congruent.

- Demonstration 2 uses the SAS Postulate to prove that two constructed triangles are congruent. Again, you may wish to have students repeat the procedure at their desks with the next triangle.

Answers
Answers appear on the teacher demonstration instructions on pages 75–76.

Geometry Activity Recording Sheet
Angle-Angle-Side Congruence
(p. 77 of this booklet)

Use With the activity on page 208 in Lesson 4-5 of the Student Edition.

Objective Use patty paper to draw and manipulate triangles to discover the Angle-Angle-Side congruence theorem.

Materials
patty paper or tracing paper
straightedge

Students can work in groups of two or three for this activity. Students draw a triangle on patty paper. Then using another piece of patty paper, students trace a side and two angles of the triangle. By tearing and reassembling the patty paper, students are lead to the conclusion that two triangles are congruent if two angles and the nonincluded side are congruent.

Answers
See Teacher Wraparound Edition page 208.

Geometry Activity Recording Sheet
Congruence in Right Triangles
(pp. 78–79 of this booklet)

Use With Lesson 4-5 as a follow-up activity. This corresponds to the activity on pages 214–215 in the Student Edition.

Objective Investigate and determine congruence that exists among right triangles.

Chapter 4 Teaching Notes and Overview

Materials
centimeter ruler*
compass*
* = available in Overhead Manipulative Resources

This activity involves three parts. In Activity 1, students examine pairs of congruent right triangles and write congruence statements using L for *leg* and H for *hypotenuse*, instead of S for *side*. In Activity 2, students use a compass and centimeter ruler to construct a right triangle with given leg and hypotenuse measures. Students determine that this is a unique triangle, and therefore SSA is a valid test of congruence in right triangles. After the two activities, students explore and prove theorems and statements regarding right triangle congruence. You may wish to select students to present their proofs for Exercises 7–11 on the chalkboard.

Answers
See Teacher Wraparound Edition pages 214–215.

Geometry Activity Recording Sheet

Isosceles Triangles
(p. 80 of this booklet)

Use With the activity on page 216 in Lesson 4-6 of the Student Edition.

Objective Use patty paper to discover that the base angles of isosceles triangles are congruent.

Materials
patty paper or tracing paper
straightedge

For this activity, students draw an acute isosceles triangle, an obtuse isosceles triangle, and a right isosceles triangle on patty paper. By folding each triangle so that the congruent sides align, students discover that the base angles are congruent. Ask students what other methods they can use to determine that the base angles are congruent.

Answers
See Teacher Wraparound Edition page 216.

 ## Mini-Project
Perimeters and Unknown Values
(p. 81 of this booklet)

Use With Lesson 4-6.

Objective Use algebra and the properties of isosceles triangles to find missing values and the perimeter of figures.

For this activity, students should work in pairs. In Exercises 1 and 2, students use algebra and the properties of isosceles triangles to find the missing values. In Exercise 3, students use algebra and the properties of isosceles triangles to find the value of *x*, and then to find the perimeter of the figure. In Exercise 4, students use algebra and congruent sides to find the perimeter of the figure.

Answers
1. 4; 9; 13
2. 26; 32
3. 73
4. 48

NAME _____ DATE _____ PERIOD ____

Geometry Activity Recording Sheet

(Use with the activity on page 179 in Lesson 4-1 of the Student Edition.)

Equilateral Triangles

Materials
patty paper or tracing paper

Analyze

1. Is $\triangle XYZ$ equilateral? Explain.

2. Use three pieces of patty paper to make a triangle that is isosceles, but not equilateral.

3. Use three pieces of patty paper to make a scalene triangle.

© Glencoe/McGraw-Hill   Teaching Geometry with Manipulatives

Geometry Activity Recording Sheet

(Use with the Lesson 4-2 Preview Activity on page 184 in the Student Edition.)

Angles of Triangles

Materials
unlined paper
straightedge
protractor

Analyze the Model
Describe the relationship between each pair.

1. $\angle A$ and $\angle DFA$
2. $\angle B$ and $\angle DFE$
3. $\angle C$ and $\angle EFC$

4. What is the sum of the measures of $\angle DFA$, $\angle DFE$, and $\angle EFC$?

5. What is the sum of the measures of $\angle A$, $\angle B$, and $\angle C$?

6. **Make a conjecture** about the sum of the measures of the angles of any triangle.

Analyze the Model

7. **Make a conjecture** about the relationship of $\angle A$, $\angle B$, and the exterior angle at C.

8. Repeat the steps for the exterior angles of $\angle A$ and $\angle B$.

9. Is your conjecture true for all exterior angles of a triangle?

10. Repeat Activity 2 with an acute triangle.

11. Repeat Activity 2 with a right triangle.

12. **Make a conjecture** about the measure of an exterior angle and the sum of the measures of its remote interior angles.

Using Overhead Manipulatives
(Use with Lesson 4-2.)

Angle Measures in Triangles

> **Objective** Discover the sum of the measures of the angles in a triangle.
>
> **Materials**
> - transparency pens*
> - scissors*
> - tape
> - straightedge
> - paper
> - blank transparency
>
> * = available in Overhead Manipulative Resources

Demonstration
Find Angle Measures of a Triangle

- Prepare a large paper triangle similar to the one shown and cut it out. Trace the triangle on a blank transparency. Label the angles on both triangles *X*, *Y*, and *Z*. (Hint: The angles of the triangle should be large enough to measure easily.)

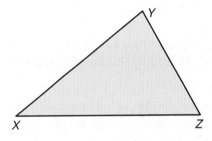

- Show students the paper triangle. Lay it on the transparency triangle to show that the triangles are congruent. Fasten the paper triangle to the transparency with a piece of tape in the middle of $\overline{XZ}$.

- Fold the paper triangle along a line parallel to $\overline{XZ}$ so vertex *Y* is on $\overline{XZ}$. Label this point *Y'* and label the endpoints of the segment formed by the fold *A* and *B* as shown.

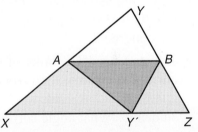

- Fold the paper triangle so that *X* and *Z* are on point *Y'* as shown.

- Ask students what is formed by ∠*XY'B* and ∠*BY'Z*. **a linear pair** Tell students that since the angles of a linear pair are supplementary, *m*∠*XY'B* + *m*∠*BY'Z* = 180.

- Ask students what allows you to say that *m*∠*XY'A* + *m*∠*AY'B* = *m*∠*XY'B*. **Angle Addition Postulate**

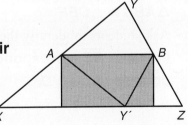

- Tell students to use substitution to find the sum of the angles of the triangle. ***m*∠*XY'A* + *m*∠*AY'B* + *m*∠*BY'Z* = 180**

- Repeat this activity with a different triangle, or have students complete the activity at their desks. Then ask what they think is true about the measures of the angles of any triangle. **The sum of the measures of the angles of a triangle is 180.**

© Glencoe/McGraw-Hill

Using Overhead Manipulatives
(Use with Lesson 4-3.)

Congruent Triangles

Objective Identify congruent triangles and corresponding parts of congruent triangles.

Materials
- scissors*
- transparency pens*
- blank transparency

* = available in Overhead Manipulative Resources

Demonstration
Identify Congruent Triangles
- If a colored transparency is available, trace, cut out, and label the triangles as shown. If you use a clear transparency, cut it in half. On each half, trace and label one of the triangles.

- Place the two triangles on the screen. Show students that the triangles can be matched exactly by placing one on top of the other and turning until all parts align. Tell students that this means that each part of the top triangle matches exactly the corresponding parts of the bottom triangle and that these triangles are congruent.

- Tell students, "The order of the letters in a congruence statement indicates the correspondence of the vertices." Ask students, "What angle corresponds to ∠A?" **∠D** Ask students, "What angle corresponds to ∠B?" **∠E** Ask students, "What angle corresponds to ∠C?" **∠F**

- Ask students to state the congruence statement for the triangles. **△ABC ≅ △DEF**

- Ask students to identify the three pairs of corresponding sides of the two triangles. **$\overline{AB}$ corresponds to $\overline{DE}$, $\overline{BC}$ corresponds to $\overline{EF}$, and $\overline{AC}$ corresponds to $\overline{DF}$.**

© Glencoe/McGraw-Hill — Teaching Geometry with Manipulatives

Using Overhead Manipulatives
(Use with Lesson 4-4.)

Tests for Congruent Triangles

Objective Use SSS and SAS postulates to test for triangle congruence.

Materials
- transparency pens*
- compass*
- straightedge
- protractor*
- 5 blank transparencies

* = available in Overhead Manipulative Resources

Demonstration 1
Use SSS to Construct Congruent Triangles
- Tell students that you want to construct a triangle with sides of lengths 5 centimeters, 7 centimeters, and 8 centimeters.
- Draw a line and label it *a*. Then choose a point on *a* and label it *X*.
- Set the compass at 8 centimeters. Place the compass point on *X* and draw an arc to intersect *a*. Label the intersection point *Y*. Tell students that *XY* = 8 centimeters.

- Set the compass at 7 centimeters. Place the compass point on *X* and draw an arc above *a*.
- Set the compass point at 5 centimeters. Place the compass point on *Y* and draw an arc to intersect the one drawn from *X*. Label the intersection point of the two arcs *Z*.

- Use the straightedge to draw $\overline{XZ}$ and $\overline{YZ}$.

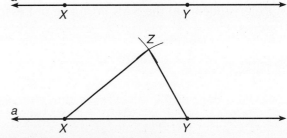

- Have students repeat the procedure with *XY* = 7 centimeters and then with *XY* = 5 centimeters. Ask them whether the triangles are congruent. **Yes, they were all made using the same compass settings.**
- Refer students to the SSS Postulate in Lesson 4-4 of *Glencoe Geometry*.

© Glencoe/McGraw-Hill

Using Overhead Manipulatives

Demonstration 2
Use SAS to Construct Congruent Triangles
- Tell students that you want to construct a triangle with two sides of lengths 6 centimeters and 9 centimeters, and the angle formed by these two sides measures 60°.
- Draw a line and label it *m*. Then choose a point on *m* and label it *R*.
- Ask a student to use a protractor to draw a 60° angle at *R* so that one side of the angle is on line *m*.

- Set the compass at 6 centimeters. Place the compass point at *R* and draw an arc to intersect the side of the 60° angle that is not on line *m*. Label the point of intersection *S*.

- Set the compass at 9 centimeters. Place the compass point at *R* and draw an arc to intersect line *m*. Label this point of intersection *T*.

- Use a straightedge to draw $\overline{ST}$.

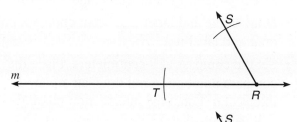

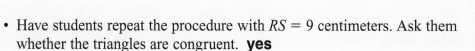

- Have students repeat the procedure with *RS* = 9 centimeters. Ask them whether the triangles are congruent. **yes**
- Refer students to the SAS Postulate in Lesson 4-4 of *Glencoe Geometry*.

© Glencoe/McGraw-Hill   Teaching Geometry with Manipulatives

NAME _____ DATE _____ PERIOD ____

Geometry Activity Recording Sheet

(Use with the activity on page 208 in Lesson 4-5 of the Student Edition.)

Angle-Angle-Side Congruence

Materials
patty paper or tracing paper
straightedge

Analyze

1. Place the original △ABC over the assembled figure. How do the triangles compare?

2. **Make a conjecture** about two triangles with two angles and the nonincluded side of one triangle congruent to two angles and the nonincluded side of the other triangle.

NAME _____ DATE _____ PERIOD _____

Geometry Activity Recording Sheet

(Use with the Lesson 4-5 Follow-Up Activity on pages 214–215 in the Student Edition.)

Congruence in Right Triangles

Materials
centimeter ruler
compass

Analyze

1. Is each pair of triangles congruent? If so, which congruence theorem or postulate applies?

2. Rewrite the congruence rules from Exercise 1 using *leg*, (L), or *hypotenuse*, (H), to replace *side*. Omit the *A* for any right angle since we know that all right triangles contain a right angle and all right angles are congruent.

Make a Conjecture

3. If you know that the corresponding legs of two right triangles are congruent, what other information do you need to declare that triangles congruent? Explain.

Make a Model

Construct right triangle *XYZ* with a hypotenuse of 10 centimeters and a leg of 7 centimeters in the space provided.

Analyze

4. Does the model yield a unique triangle?

5. Can you use the lengths of the hypotenuse and a leg to show right triangles are congruent?

6. **Make a conjecture** about the case of SSA that exists for right triangles.

© Glencoe/McGraw-Hill   78   Teaching Geometry with Manipulatives

Geometry Activity Recording Sheet

Proof
Write a paragraph proof of each theorem.

7. Theorem 4.6

8. Theorem 4.7

9. Theorem 4.8 (*Hint*: There are two possible cases.)

Use the figure to write a two-column proof.

10. Given: $\overline{ML} \perp \overline{MK}, \overline{JK} \perp \overline{KM}$
    $\angle J \cong \angle L$
    Prove: $\overline{JM} \cong \overline{KL}$

11. Given: $\angle J$ and $\angle L$ are rt. $\angle$s.
    $\overline{ML} \parallel \overline{JK}$
    Prove: $\overline{ML} \cong \overline{JK}$

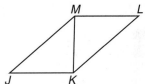

NAME _____ DATE _____ PERIOD _____

Geometry Activity Recording Sheet

(Use with the activity on page 216 in Lesson 4-6 of the Student Edition.)

Isosceles Triangles

Materials
patty paper or tracing paper
straightedge

Analyze

1. What do you observe about $\angle A$ and $\angle B$?

2. Draw an obtuse isosceles triangle. Compare the base angles.

3. Draw a right isosceles triangle. Compare the base angles.

© Glencoe/McGraw-Hill   Teaching Geometry with Manipulatives

NAME _____ DATE _____ PERIOD ____

Mini-Project

(Use with Lesson 4-6.)

Perimeters and Unknown Values

Work with a partner and discuss how to use the given information to find the unknown values in each of the following. Then find the values.

1.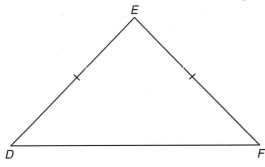

   $DF = 3x + 1$, $DE = x + 5$
   The perimeter is 31.
   Find x, DE, and DF.

2.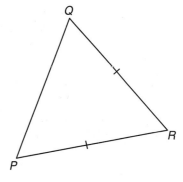

   $PQ = x + 3$, $QR = x + 6$
   The perimeter is 93.
   Find x and PR.

3.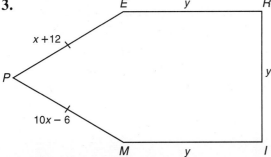

   $y = 5x + 5$
   Find the perimeter of the figure.

4.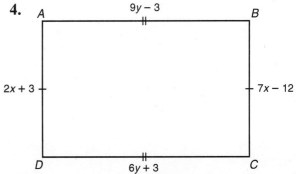

   Find the perimeter of the figure.

© Glencoe/McGraw-Hill   81   Teaching Geometry with Manipulatives

Chapter 5
Relationships in Triangles
Teaching Notes and Overview

Geometry Activity Recording Sheet
Bisectors, Medians, and Altitudes
(pp. 87–88 of this booklet)

Use With Lesson 5-1 as a preview activity. This corresponds to the activity on pages 236–237 in the Student Edition.

Objective Construct perpendicular bisectors, medians, angle bisectors, and altitudes of triangles.

Materials
compass*
straightedge
* = available in Overhead Manipulative Resources

This activity involves four types of constructions: perpendicular bisectors, medians, angle bisectors, and altitudes. In the Student Edition, each construction is explained with one segment or angle. You may wish to perform these constructions on the overhead while students are performing the construction at their desks. Then have students complete the other two constructions for each triangle on their own. Once all of the constructions are completed, discuss with students any relationships they noticed among bisectors, medians, and altitudes.

Answers
See Teacher Wraparound Edition pages 236–237.

Using Overhead Manipulatives
Constructing a Median in a Triangle
(p. 89 of this booklet)

Use With Lesson 5-1.

Objective Construct medians of a triangle.

Materials
transparency pens*
compass*
straightedge
blank transparency
* = available in Overhead Manipulative Resources

This demonstration involves using a compass and straightedge to construct a median of a triangle. Before the extension, you may wish to have two students come to the overhead and construct the other two medians of the triangle.

In this extension, students are to construct the medians of an acute triangle, obtuse triangle, and right triangle at their desks. To save on time, you may wish to divide the class into thirds. Have each group construct the medians of one type of triangle. Then have students share their findings with the class.

Answers
Answers appear on the teacher demonstration instructions on page 89.

Using Overhead Manipulatives
Investigating Perpendicular Bisectors
(p. 90 of this booklet)

Use With Lesson 5-1.

Objective Use paper folding to find the perpendicular bisectors of a triangle and investigate the relationships between the bisectors.

Materials
transparency pens*
protractor*
blank transparency
* = available in Overhead Manipulative Resources

© Glencoe/McGraw-Hill 83 Teaching Geometry with Manipulatives

Chapter 5 Teaching Notes and Overview

This demonstration involves paper folding as a way to discover that the perpendicular bisectors of a triangle intersect at one point. While you are completing the paper folding at the overhead, you may wish to have students complete the activity at their desk with a blank sheet of paper.

In this extension, students use the same technique to find the intersection of the perpendicular bisectors of a right triangle and an obtuse triangle. Before students complete the extension, you may wish to have students make a conjecture about the intersection of the perpendicular bisectors of these two types of triangles.

Answers
Answers appear on the teacher demonstration instructions on page 90.

Geometry Activity
Special Segments in a Triangle
(pp. 91–92 of this booklet)

Use With Lesson 5-1.

Objective Identify and apply the definitions of angle bisectors, altitudes, and medians to congruent triangles.

Materials
classroom set of Geometry Activity worksheets
patty paper or waxed paper
construction paper

Have students begin by completing Exercises 1 and 2. While discussing the triangles drawn by students, you may wish to select students to draw the triangles they drew on the chalkboard or overhead.

Give each student two pieces of patty paper to use with Exercises 3 and 5. Have students complete the remaining exercises. For Exercise 6, you may wish to allow students to work in pairs. Tell students it may be helpful to use colored pencils to outline each different pair of congruent triangles.

Give each student a piece of construction paper. Have them cut out a triangle of any size. Instruct them to fold one side of the triangle so that the two vertices meet, and pinch a crease on the midpoint. Fold a crease from the opposite vertex to the midpoint just found. Tell students that this segment is one median of the triangle. Have students repeat this process to find the other two medians of the triangle. Have students balance the triangle on a pencil point that is placed at the point of intersection of the medians. Ask students what this point of intersection is called. Explain that the centroid is the center of gravity of any triangle.

Answers

1a. Sample answer:

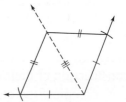

1b. Sample answer:

2a. Sample answer:

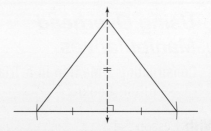

2b. Sample answer:

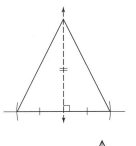

3.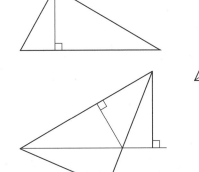

4. congruent (HL); nothing; nothing

5. Answers may vary.

6. $\triangle AXC \cong \triangle AXB \cong \triangle BYC \cong \triangle BYA \cong \triangle CZA \cong \triangle CZB$; $\triangle BQZ \cong \triangle CQY$; $\triangle BXQ \cong \triangle CXQ$; $\triangle AZQ \cong \triangle AYQ$; $\triangle AQC \cong \triangle AQB$

Mini-Project
Folding Triangles
(p. 93 of this booklet)

Use With Lesson 5-1.

Objective Copy, cut out, fold, and draw triangles that represent the described segments.

Materials
unlined paper
straightedge
scissors*

* = available in Overhead Manipulative Resources

For this activity, students should work in groups of two or three. Students should first copy, cut out, and label each of the four triangles. Then students should fold and make a drawing of each triangle that represents the described altitude, perpendicular bisector, angle bisector, or median.

Answers

1.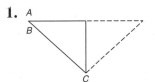

2.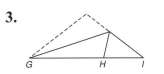

3.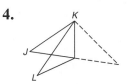

4.

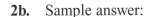
Geometry Activity Recording Sheet
Inequalities for Sides and Angles of Triangles
(p. 94 of this booklet)

Use With the activity on page 249 in Lesson 5-2 of the Student Edition.

Objective Draw and measure the sides and angles of a triangle to discover Theorem 5.9.

Materials
ruler*
protractor*

* = available in Overhead Manipulative Resources

Chapter 5 Teaching Notes and Overview

For this activity, students can work in pairs. Students draw an acute scalene triangle and measure its sides and angles. By examining these measurements, students discover that the angle opposite the longest side of a triangle is the largest angle, and that the angle opposite the shortest side of a triangle is the smallest angle. Before reviewing students' answers to Exercises 1–4, you may wish to select a few students to write their side measures and angle measures on the chalkboard or overhead.

Answers
See Teacher Wraparound Edition page 249.

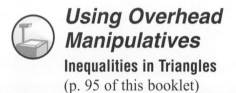

Using Overhead Manipulatives
Inequalities in Triangles
(p. 95 of this booklet)

Use With Lesson 5-4.

Objective Discover the relationships between sides and angles in a triangle.

Materials
lined paper transparency*
transparency pens*
protractor*
blank transparency
* = available in Overhead Manipulative Resources

This demonstration engages students in a hands-on method to discover the relationship among the sides and angles of a triangle. You may wish to choose one member from each of the three groups to write their side and angle measures on the chalkboard or overhead. This will aid in your discussion of the relationship between the sides and angles of a triangle.

Answers
Answers appear on the teacher demonstration instructions on page 95.

Using Overhead Manipulatives
Investigating the Triangle Inequality Theorem
(pp. 96–97 of this booklet)

Use With Lesson 5-4.

Objective Investigate the Triangle Inequality Theorem.

Materials
pipe cleaners or twist ties
centimeter ruler*
transparency pens*
lined paper transparency*
blank transparency
* = available in Overhead Manipulative Resources

This activity includes two demonstrations to investigate the Triangle Inequality Theorem.

- For Demonstration 1 make sure you have enough pipe cleaners or twist ties for students to complete the activity at their desks. This demonstration involves using a pipe cleaner or twist tie to discover lengths of sides that form triangles and lengths of sides that do not form triangles. Once students create their lists of triangles, they add the measures of the two shorter sides to discover that they must be greater than the third side in order to form a triangle.

- Demonstration 2 provides an example of how to find the range of the length of the third side of a triangle, given the lengths of two sides. After you complete the demonstration, you may wish to give students the lengths of two sides of a triangle and see if they can find the range for the length of the third side.

Answers
Answers appear on the teacher demonstration instructions on pages 96–97.

© Glencoe/McGraw-Hill   Teaching Geometry with Manipulatives

NAME _____ DATE _____ PERIOD ____

Geometry Activity Recording Sheet

(Use with the Lesson 5-1 Preview Activity on pages 236–237 in the Student Edition.)

Bisectors, Medians, and Altitudes

Materials
compass
straightedge

Construction 1

Construct the perpendicular bisector of $\overline{AC}$.

1. Construct the perpendicular bisectors for the other two sides.

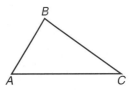

2. What do you notice about the intersection of the perpendicular bisectors?

Construction 2

Construct the median of $\overline{BC}$.

3. Construct the median of the other two sides.

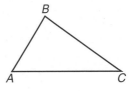

4. What do you notice about the medians of a triangle?

Construction 3

Construct the altitude of $\overline{AC}$.

5. Construct the altitudes to the other two sides. (*Hint:* You may need to extend the lines containing the sides of your triangle.)

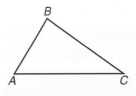

6. What observation can you make about the altitudes of your triangle?

Construction 4

Construct the angle bisector of $\angle A$.

7. Construct the angle bisectors for the other two angles.

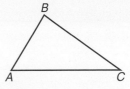

8. What do you notice about the angle bisectors?

Geometry Activity Recording Sheet

Analyze

9. Repeat the four constructions for each type of triangle. You may need to use a separate piece of paper or the back of this page.

   a. obtuse scalene

   b. right scalene

   c. isosceles

   d. equilateral

Make a Conjecture

10. Where do the lines intersect for acute, obtuse, and right triangles?

11. Under what circumstances do the special lines of triangles coincide with each other?

Using Overhead Manipulatives
(Use with Lesson 5-1.)

Constructing a Median in a Triangle

Objective Construct medians of a triangle.

Materials
- transparency pens*
- compass*
- straightedge
- blank transparency

* = available in Overhead Manipulative Resources

Demonstration
Construct Medians of a Triangle
- Draw a large triangle on a blank transparency. Label the triangle *ABC*.
- Open the compass to a setting that is longer than half the length of $\overline{AB}$. Place the compass at point *A*. Draw a small arc above and below $\overline{AB}$.

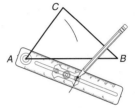

- Show students that you are keeping the same compass setting. Place the compass at point *B* and draw arcs that intersect the arcs you made from point *A*.

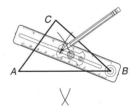

- Draw a segment connecting the points where the arcs intersect. This segment intersects $\overline{AB}$. Point out that this is the perpendicular bisector of $\overline{AB}$. Label the point of intersection point *M*.

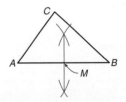

- Draw $\overline{MC}$. Tell students that $\overline{MC}$ is a median of $\triangle ABC$.

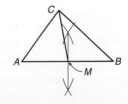

Extension
Construct Medians of an Acute Triangle
- Have students draw an acute triangle on a sheet of paper and construct all three medians. Ask them to describe any relationship they see as a result. **They all intersect at one point.** Have students construct the medians of a right triangle and of an obtuse triangle. Ask "Are the results the same?" **yes**

© Glencoe/McGraw-Hill 89 Teaching Geometry with Manipulatives

Using Overhead Manipulatives
(Use with Lesson 5-1.)

Investigating Perpendicular Bisectors

Objective Use paper folding to find the perpendicular bisectors of a triangle and investigate the relationships between the bisectors.

Materials
- transparency pens*
- protractor*
- blank transparency

* = available in Overhead Manipulative Resources

Demonstration
Find Perpendicular Bisectors of a Triangle
- Draw an acute triangle *DEF* on a blank transparency. Fold the transparency so that one vertex falls on a second vertex as shown.

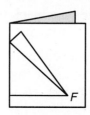

- Unfold the transparency and use a straightedge to draw a line on the fold.

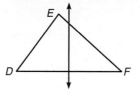

- Repeat the procedure with the other two pairs of vertices.
- Tell students that you have constructed the three perpendicular bisectors of the triangle. Ask students what they notice about the perpendicular bisectors. **They all intersect in one point.**

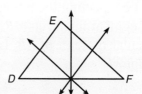

Extension
Find Perpendicular Bisectors of a Right Triangle and an Obtuse Triangle
- Have students draw a right triangle and an obtuse triangle on a piece of unlined or waxed paper. Then use the paper-folding technique to find the perpendicular bisectors. Is the result the same as for the acute triangle? **In the right triangle, the three lines intersect in a point on the hypotenuse. In the obtuse triangle, the lines intersect in a point outside of the triangle.**

NAME _____ DATE _____ PERIOD ____

Geometry Activity

(Use with Lesson 5-1.)

Special Segments in a Triangle

1. Using the dotted lines in each figure as angle bisectors, construct pairs of congruent triangles. Identify the rule for proving the triangles are congruent.

   Example: 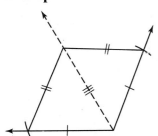   **a.** 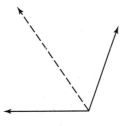   **b.**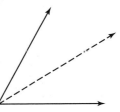

2. The dotted lines represent perpendicular bisectors of the segments. Draw segments to construct sets of congruent triangles. Then state the rule for proving the triangles congruent.

   Example: 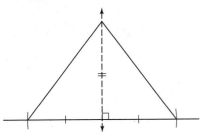   **a.**    **b.**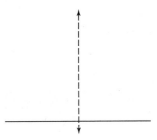

3. Trace each triangle below onto waxed paper. Then make a crease along the altitude to each side. Sketch your waxed paper altitude lines on the triangles below.

   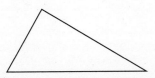      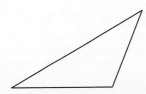

© Glencoe/McGraw-Hill   91   Teaching Geometry with Manipulatives

Geometry Activity

4. Given: △ABC is isosceles with legs $\overline{AB}$ and $\overline{AC}$. What can be said about the two triangles formed by drawing an altitude from

vertex A? _____

vertex B? _____

vertex C? _____

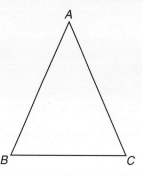

5. Construct a right triangle using the paper-folding techniques in this activity. Write instructions for your construction.

_____
_____
_____

6. In equilateral △ABC below, $\overrightarrow{AX}$, $\overline{BY}$, and $\overline{CZ}$ are altitudes. List all pairs of congruent triangles.

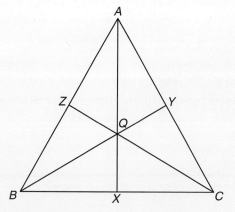

_____
_____

NAME _____ DATE _____ PERIOD _____

Mini-Project

(Use with Lesson 5-1.)

Folding Triangles

Cut out a copy of each triangle. Label the vertices on the interior of the angle. Fold it so that the fold represents the segment described. Then make a drawing of the folded triangle.

Example: the altitude of $\triangle PQR$ from vertex P

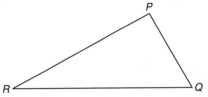

Fold the triangle through vertex P and perpendicular to $\overline{RQ}$. The folded side is the altitude of $\triangle PQR$ from P.

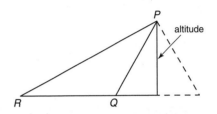

1. the altitude of isosceles $\triangle ABC$ through vertex angle C

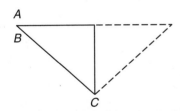

2. the perpendicular bisector of $\overline{DF}$

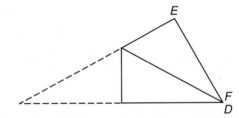

3. the bisector of angle G

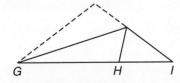

4. the median through vertex K

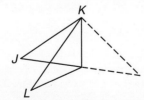

© Glencoe/McGraw-Hill   Teaching Geometry with Manipulatives

NAME _____ DATE _____ PERIOD ____

Geometry Activity Recording Sheet

(Use with the activity on page 249 in Lesson 5-2 of the Student Edition.)

Inequalities for Sides and Angles of Triangles

Materials
ruler
protractor

Model
Draw an acute scalene triangle in the space provided at the right. Record the side measures and angle measures in the tables.

Side	Measure
$\overline{BC}$	
$\overline{AC}$	
$\overline{AB}$	

Angle	Measure
$\angle A$	
$\angle B$	
$\angle C$	

Analyze

1. Describe the measure of the angle opposite the longest side in terms of the other angles.

2. Describe the measure of the angle opposite the shortest side in terms of the other angles.

3. Repeat the activity using other triangles.

Make a Conjecture

4. What can you conclude about the relationship between the measures of sides and angles of a triangle?

© Glencoe/McGraw-Hill   Teaching Geometry with Manipulatives

Using Overhead Manipulatives
(Use with Lesson 5-4.)

Inequalities in Triangles

Objective Discover the relationships between sides and angles in a triangle.

Materials
- lined paper transparency*
- transparency pens*
- protractor*
- blank transparency

* = available in Overhead Manipulative Resources

Demonstration
Discover Inequality Relationships in Triangles
- Draw a scalene triangle ABC on a blank transparency.
- Ask a student to measure each angle of the triangle with the protractor and write the angle measures on the transparency. (Sample measures are given.)

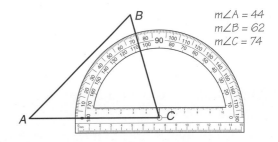

- Using the ruler edge of the protractor, have a student measure the length of each side of the triangle and write the lengths of the sides of the triangle on the transparency also. (Sample lengths are given.)

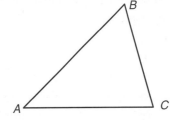

- List the angles and the sides in order from greatest to least. $\angle C, \angle B, \angle A$; $\overline{AB}, \overline{AC}, \overline{BC}$
- Divide the class into three groups. Have students in the first group draw an acute scalene triangle, students in the second group draw a right scalene triangle, and students in the third group draw an obtuse scalene triangle. Each student should label his or her triangle ABC and measure each angle and each side. Ask students to list the angles and sides in order from least to greatest.
- On the lined paper transparency, write the list of sides and angles from the triangle you drew on the first transparency. Ask volunteers to give the order of sides and angles found in their triangle until all of the orders are listed. Do students see a pattern? **The order of the angles corresponds to the order of the opposite sides.**

© Glencoe/McGraw-Hill   95   Teaching Geometry with Manipulatives

Using Overhead Manipulatives
(Use with Lesson 5-4.)

Investigating the Triangle Inequality Theorem

Objective Investigate the triangle inequality theorem.

Materials
- pipe cleaners or twist ties
- centimeter ruler*
- transparency pens*
- lined paper transparency*
- blank transparency

* = available in Overhead Manipulative Resources

Demonstration 1
Use Pipe Cleaners or Twist Ties to Investigate the Triangle Inequality Theorem

- Place a pipe cleaner on a blank transparency on the overhead projector. Then choose two points on the pipe cleaner and attempt to bend the pipe cleaner to form a triangle.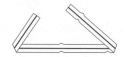

- Use a centimeter ruler to measure each segment of the pipe cleaner to the nearest tenth of a centimeter. Record the measurements in order from least to greatest on the transparency.

- Divide students into groups of four or five. Distribute pipe cleaners and have students perform the investigation several times. Groups should divide a piece of paper into two columns and list the combinations of measures that yield triangles on one side and the combinations that do not yield triangles on the other side. Each combination of measures should be written in order from least to greatest.

- After the groups have made a list of 15 combinations of measures, have students add the two lessor measures in each combination and compare the sum to the greatest measure. Ask students how the sum compares to the greatest measure in combinations that do not form a triangle. **The sum is less than or equal to the greatest measure.** Ask students how the sum compares to the greatest measure in combinations that do form a triangle. **The sum is greater than the greatest measure.**

Demonstration 2
Use Pipe Cleaners or Twist Ties to Investigate the Triangle Inequality Theorem

- Join two pipe cleaners at one end so that one measures 6 centimeters and the other measures 10 centimeters. Explain to students that these two segments represent two sides of a triangle.

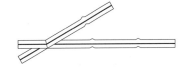

- Move the two segments together slowly, explaining that the third side of the triangle is the segment between the endpoints of the segments. When the segments overlap, the figure becomes a segment. Measure the length from the end of the shorter pipe cleaner to the end of the longer pipe cleaner. **4 cm** Explain that the length of the third side of the triangle must be greater than 4 centimeters.

- Move the two segments apart again. When the two segments are stretched out to make a straight line, the figure becomes a segment again. Measure the segment from end to end. **16 cm** Explain that the length of the third side must be less than 16 centimeters.

- Tell students, "If the third side of the triangle is x centimeters long, we can write the inequality $4 < x < 16$." Ask students to check values of x in the ranges $x \leq 4$, $4 < x < 16$, and $x \geq 16$ with the triangle inequality. Ask, "Does the solution check?" **yes**

Chapter 6
Proportions and Similarity
Teaching Notes and Overview

Geometry Activity Recording Sheet
Similar Triangles
(p. 102 of this booklet)

Use With the activity on page 298 in Lesson 6-3 of the Student Edition.

Objective Draw two triangles with given measurements, and calculate corresponding side ratios to determine if the triangles are similar.

Materials
centimeter ruler*
protractor*
* = available in Overhead Manipulative Resources

For this activity, students can work in pairs. Students draw two triangles with given measurements. By measuring the sides and calculating the ratio of corresponding sides, students discover that the triangles are similar. This leads to the Angle-Angle similarity postulate.

Answers
See Teacher Wraparound Edition page 298.

Using Overhead Manipulatives
Similar Triangles
(p. 103 of this booklet)

Use With Lesson 6-3.

Objective Investigate the relationships between the measures of similar triangles.

Materials
centimeter ruler*
protractor*
transparency pens*
blank transparency
* = available in Overhead Manipulative Resources

This demonstration involves construction, measurement, and ratios in discovering the Angle-Angle similarity postulate. You may wish to have students find the ratio of corresponding sides of the similar triangles. Ask students how these three ratios compare.

Answers
Answers appear on the teacher demonstration instructions on page 103.

Geometry Activity
Similar Triangles
(pp. 104–106 of this booklet)

Use With Lesson 6-3.

Objective Determine measurements of sides and angles of similar triangles using indirect measurement techniques.

Materials
classroom set of Geometry Activity worksheets
classroom set of mirrors, approximately 3-inch square
transparency master of Geometry Activity
blank transparency

You may wish to complete the following demonstration before handing out the worksheet.

Demonstration:
Display the transparency master of 30°-60°-90° triangles on the overhead. On a blank transparency, draw more 30°-60°-90° triangles of different sizes. Overlap the triangles in various ways and discuss the characteristics of similar triangles. For each pair of similar triangles you illustrate, overlap the angles to show they are congruent.

© Glencoe/McGraw-Hill 99 Teaching Geometry with Manipulatives

Chapter 6 Teaching Notes and Overview

Model Figures 1-3 one at a time on the overhead. Label the vertices of each triangle. Have students identify the corresponding angles of similar triangles.

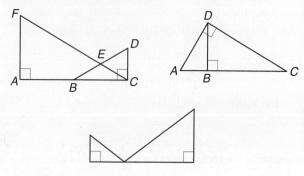

Hand out the worksheets and have students examine the diagram. Discuss the corresponding angles of the similar triangles. Explain which angle is the angle of incidence and which is the angle of reflection. Explain that the angles are congruent because the angle of incidence is equal to the angle of reflection of the mirror. Discuss the instructions for the indirect measurement activity. Then have students complete the activity. Once students have completed the activity, you may wish to have students compare their measurements and answers.

Answers

1–8. See students' work.

9. far away from the mirror; closer to the object: angle became greater; farther from object: angle became smaller

10. very tall: angle became very small, very short: angle became very great

11. No; you must be able to adjust to see the top of the object.

Mini-Project
Measuring Height
(p. 107 of this booklet)

Use With Lesson 6-3.

Objective Make a hypsometer to measure the heights of various objects around your school.

Materials

heavy piece of cardboard at least 10 cm by 20 cm
straw
string about 25 cm long
small weight

For this activity, students should work in groups of three or four. Students begin by making a hypsometer using the cardboard, straw, string, and weight. Once the hypsometer is complete, students use it to measure the heights of various objects around your school. The height of each object is found by using similar triangles that are formed by the hypsometer and each object. Inform students that drawing an illustration of the similar triangles may be helpful in finding the height of each object.

Answers

1–5. See students' work.

Using Overhead Manipulatives
Trisecting a Segment
(p. 108 of this booklet)

Use With Lesson 6-4.

Objective Trisect a segment.

© Glencoe/McGraw-Hill

Teaching Geometry with Manipulatives

Chapter 6 Teaching Notes and Overview

Materials
compass*
straightedge
transparency pens*
blank transparency

* = available in Overhead Manipulative Resources

This demonstration involves using a compass and straightedge to trisect a segment. Once you complete the construction at the overhead, you may wish to have a student come to the overhead and use a ruler to verify that each of the three segments are congruent.

In this extension, students divide a segment into four congruent parts by construction. Have students use a ruler to verify that the segments are congruent.

Answers
Answers appear on the teacher demonstration instructions on page 108.

Geometry Activity Recording Sheet
Sierpinski Triangle
(p. 109 of this booklet)

Use With Lesson 6-6 as a preview activity. This corresponds to the activity on page 324 in the Student Edition.

Objective Draw and investigate patterns in the Sierpinski Triangle.

Materials
isometric dot paper
straightedge

In this activity, students use isometric dot paper to draw Stages 0–4 of the Sierpinski Triangle. Students examine patterns in the perimeter and similarity at various stages of the Sierpinski Triangle. You may wish to draw the first few stages at the overhead on an isometric dot paper transparency. Ask students why it is helpful to use dot paper to draw the Sierpinski Triangle.

Answers
See Teacher Wraparound Edition page 324.

© Glencoe/McGraw-Hill 101 Teaching Geometry with Manipulatives

NAME _____ DATE _____ PERIOD ____

Geometry Activity Recording Sheet

(Use with the activity on page 298 in Lesson 6-3 of the Student Edition.)

Similar Triangles

Materials
centimeter ruler
protractor

Collect Data
Draw $\triangle DEF$ and $\triangle RST$ in the space provided at the right. Record the side measures in the table.

Side	Measure
$\overline{EF}$	
$\overline{ED}$	
$\overline{RS}$	
$\overline{RT}$	

Calculate each ratio.

$\dfrac{FD}{ST} =$ $\dfrac{EF}{RS} =$ $\dfrac{ED}{RT} =$

Analyze the Data

1. What can you conclude about all of the ratios?

2. Repeat the activity with two more triangles with the same angle measures, but different side measures. Then repeat the activity with a third pair of triangles. Are all of the triangles similar? Explain.

3. What are the minimum requirements for two triangles to be similar?

© Glencoe/McGraw-Hill   Teaching Geometry with Manipulatives

Using Overhead Manipulatives
(Use with Lesson 6-3.)

Similar Triangles

Objective Investigate the relationships between the measures of similar triangles.

Materials
- centimeter ruler*
- protractor*
- transparency pens*
- blank transparency

* = available in Overhead Manipulative Resources

Demonstration
Investigate Similar Triangles

- Use a protractor and a centimeter ruler to draw △RST with $m\angle T = 40$, $RT = 10$ centimeters, and $m\angle R = 65$.

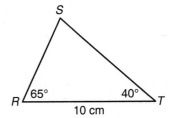

- Measure $\overline{SR}$. Then ask a student to find the value of $\frac{RT}{SR}$ with a calculator. **about 1.50**
- Draw △XYZ with $m\angle Z = 40$, $XZ = 6$ centimeters, and $m\angle X = 65$.

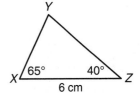

- Measure $\overline{YX}$ and ask a student to find $\frac{XZ}{YX}$. **about 1.5**
  Ask students what they observe about these ratios. **They are the same for the two triangles.**
- Ask students to calculate $\frac{SR}{ST}$ and $\frac{YX}{YZ}$. Are they equal also? **about 0.71; yes**
- Ask, "Are $\frac{TR}{ST}$ and $\frac{ZX}{YZ}$ equal?" **about 1.07; yes**
- Ask, "How are $\overline{TR}$ and $\overline{ZX}$, and $\overline{SR}$ and $\overline{YX}$ related?" **They are corresponding sides of the triangles.**
- Divide students into groups of three or four. Have each group choose measures for $\angle A$ and $\angle B$ of △ABC. Each student in the group should choose a different measure for $\overline{AB}$ and draw a triangle. Then have each student find $\frac{AB}{AC}$, $\frac{AB}{BC}$, and $\frac{BC}{AC}$ and compare his or her ratios with those of other members of the group. **The ratios are all the same for the same pairs of sides.** Have students make a conjecture about two triangles that have two pairs of angles congruent.

NAME _____ DATE _____ PERIOD ____

Geometry Activity Transparency Master

(Use with Lesson 6-3.)

Similar Triangles

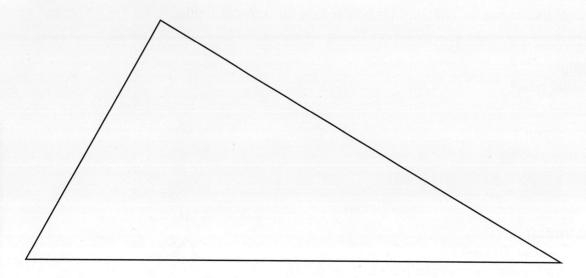

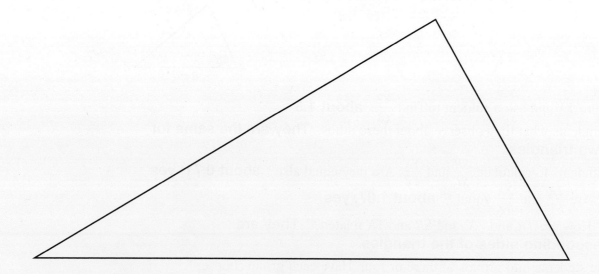

© Glencoe/McGraw-Hill   104   Teaching Geometry with Manipulatives

Geometry Activity

(Use with Lesson 6-3.)

Similar Triangles

Measuring Instructions:

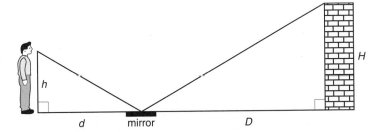

1. Place a mirror on a flat surface.

2. Stand erect. Do not lean over to view the mirror. While looking into the mirror, step forward or backward to view the top of the object.

3. Determine distance d and D. Use your pace, or step, length as an estimate. Then find your height (eye level).

4. Use the proportion $\frac{H}{D} = \frac{h}{d}$ to calculate H.

Use indirect measurement to find each of the following heights. (Draw a diagram for each problem.)

1. the school building

2. a tree

3. a goalpost

4. an electric pole

5. a sign

6. a flagpole

Geometry Activity

Choose your own objects for Exercises 7 and 8.

7. _____   8. _____

9. When measuring very tall objects, did you stand far away from or close to the mirror?

   _____

   How did the angle of incidence change as you moved the mirror closer to or farther from the object you measured?

   _____

10. What happened to your angle of incidence/reflection as your objects were very tall or very short?

    _____

11. Could all heights be measured using the same angle of incidence/reflection?

    _____

    Why or why not?

NAME _____ DATE _____ PERIOD ____

Mini-Project

(Use with Lesson 6-3.)

Measuring Height

A hypsometer can be used to measure the height of an object. To construct your own hypsometer, you will need a heavy piece of cardboard at least 10 centimeters by 20 centimeters in dimension, a straw, a string about 25 centimeters long, and a small weight.

Mark off 1 centimeter increments on each of the 10-centimeter and 20-centimeter sides of the cardboard. Attach the straw to one of the 20-centimeter sides. Then attach the weight to one end of the string, and attach the other end of the string to one end of the straw, as shown in the figure below.

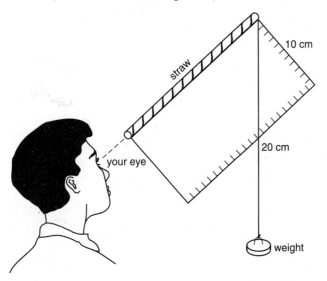

Sight through the straw to the top of the object. Note where the free-hanging string crosses the scale. Use similar triangles to find the height of the object.

Use your hypsometer to find the height of each of the following.

1. your school's flagpole

2. any tree on your school's property

3. the highest point on your school building

4. the goalposts on the football field

5. the hoop on the basketball court

© Glencoe/McGraw-Hill   Teaching Geometry with Manipulatives

Using Overhead Manipulatives
(Use with Lesson 6-4.)

Trisecting a Segment

Objective Trisect a segment.

Materials
- compass*
- straightedge
- transparency pens*
- blank transparency

* = available in Overhead Manipulative Resources

Demonstration
Trisect a Segment
- Draw a segment ST to be trisected. Then draw a ray SV.

- With the compass point at S, mark an arc on $\overrightarrow{SV}$ at A. Then use the same compass setting to construct $\overline{AB}$ and $\overline{BC}$ so that $\overline{SA} \cong \overline{AB} \cong \overline{BC}$.

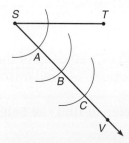

- Draw $\overline{CT}$. Then construct lines through B and A that are parallel to $\overline{CT}$. You may wish to refer students to the construction of a parallel line through a point not on the line in Lesson 3-5 of *Glencoe Geometry*. Call the points of intersection Q and R.
- Ask students why we can say that $\overline{SQ} \cong \overline{QR} \cong \overline{RT}$. **If parallel lines cut off congruent segments on one transversal, then they cut off congruent segments on every transversal.**

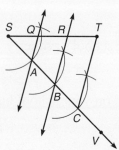

Extension
Divide a Segment into Four Congruent Parts
- Have students work in pairs to divide a segment into four congruent parts using the technique you demonstrated.

NAME _____ DATE _____ PERIOD ___

Geometry Activity Recording Sheet

(Use with the Lesson 6-6 Follow-Up Activity on page 324 in the Student Edition.)

Sierpinski Triangle

Materials
isometric dot paper
straightedge

Analyze the Data

1. Continue the process through Stage 4. How many nonshaded triangles do you have at Stage 4?

2. What is the perimeter of a nonshaded triangle in Stage 1 through Stage 4?

3. If you continue the process indefinitely, describe what will happen to the perimeter of each nonshaded triangle.

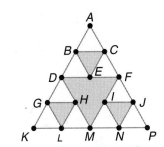

4. Study △DFM in Stage 2 of the Sierpenski Triangle shown at the right. Is this an equilateral triangle? Are △BCE, △GHL, or △IJN equilateral?

5. Is △BCE ~ △DFM? Explain your answer.

6. How many Stage 1 Sierpinski triangles are there in Stage 2?

Make a Conjecture

7. How can three copies of a Stage 2 triangle be combined to form a Stage 3 triangle?

8. Combine three copies of the Stage 4 Sierpinski triangle. Which stage of the Sierpinski Triangle is this?

9. How many copies of the Stage 4 triangle would you need to make a Stage 6 triangle?

Chapter 7 Right Triangles and Trigonometry
Teaching Notes and Overview

Mini-Project
The Pythagorean Theorem
(p. 114 of this booklet)

Use With Lesson 7-2.

Objective Trace, cut out, and rearrange the six parts of the square to create an area model of the Pythagorean theorem.

Materials
unlined paper
straightedge
scissors*

* = available in Overhead Manipulative Resources

For this activity, students should work in pairs. Students begin by tracing and cutting out the six parts of the large square. Remind students to number each part. The students then rearrange the parts to form two smaller squares. By determining the area of the large square and the two smaller squares, the students discover another method to proving the Pythagorean theorem.

Answers
1. c^2
2.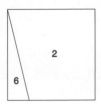
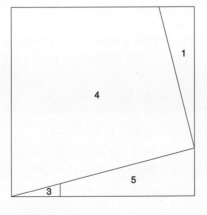

3. a^2; b^2
4. sum of areas of two smaller squares = area of larger square
5. yes

Geometry Activity Recording Sheet
The Pythagorean Theorem
(p. 115 of this booklet)

Use With Lesson 7-2 as a preview activity. This corresponds to the activity on page 349 in the Student Edition.

Objective Use patty paper to discover the Pythagorean Theorem as it was used by ancient cultures.

Materials
patty paper or square pieces of tracing paper
ruler*

* = available in Overhead Manipulative Resources

For this activity, students can work in groups of two or three. Students draw and label the area of different regions on two pieces of patty paper. By shading equal areas on the two pieces of patty paper, students discover that $a^2 + b^2 = c^2$. Students then use a ruler to measure a, b, and c, and verify that $a^2 + b^2 = c^2$.

Answers
See Teacher Wraparound Edition page 349.

Chapter 7 Teaching Notes and Overview

Using Overhead Manipulatives

The Pythagorean Theorem
(p. 116 of this booklet)

Use With Lesson 7-2.

Objective Find the relationship among the sides of a right triangle.

Materials
centimeter grids transparency*
straightedge
transparency pens*
scissors*
2 blank transparencies
* = available in Overhead Manipulative Resources

This demonstration is an area model of the Pythagorean Theorem. Show students how centimeter grid paper can be used as a ruler to find the measure of the hypotenuse. Ask students for other methods that could have been used to find the length of the hypotenuse.

In this extension, students explore the same technique with right, acute, and obtuse triangles. You may wish to have students work in groups to complete the extension.

Answers
Answers appear on the teacher demonstration instructions on page 116.

Geometry Activity Recording Sheet

Trigonometric Ratios
(p. 117 of this booklet)

Use With the activity on page 365 in Lesson 7-4 of the Student Edition.

Objective Study patterns in similar right triangles to discover that trigonometric ratios only depend on the measures of the angles.

Materials
lined paper
scissors*
centimeter ruler*
* = available in Overhead Manipulative Resources

For this activity, students use a piece of unlined paper to cut out and draw several similar triangles. Students then measure side lengths and calculate the sine, cosine, and tangent of $\angle A$ in each triangle. By observing the same ratios, students discover that trigonometric ratios depend only on angle measures.

Answers
See Teacher Wraparound Edition page 365.

Geometry Activity

Trigonometry
(pp. 118–120 of this booklet)

Use With Lesson 7-4.

Objective Use indirect measurement as an application of trigonometric functions.

Materials
classroom set of Geometry Activity worksheets
transparency master of Geometry Activity
transparent tape
kite string
protractor*, 5" × 7" index card, straw, paper clip, and scientific calculator for each pair of students
* = available in Overhead Manipulative Resources

Prior to activity, assemble a model of a hypsometer according to the instructions on the transparency master.

© Glencoe/McGraw-Hill   Teaching Geometry with Manipulatives

Display the transparency master on the overhead and have pairs of students assemble a hypsometer of their own. Show students the hypsometer you made prior to class. If protractors are not available, you can use copies of the protractor model at the bottom of the transparency master.

Have students practice finding a horizontal line of sight and reading the hypsometer angle of inclination. Once students are comfortable using the hypsometer, have them complete the activity. You may wish to have students compare measurements and answers after they complete the activity.

Answers

1. ≈ 15.5 ft
2. ≈ 19.3 m

3–10. See students' work.

Geometry Activity Recording Sheet

Trigonometric Identities
(p. 121 of this booklet)

Use With Lesson 7-7 as a follow-up activity. This corresponds to the activity on page 391 in the Student Edition.

Objective Explore and verify trigonometric identities.

Materials
none

In this activity, students learn about trigonometric identities. You may wish to discuss Activity 1 and Activity 2 as a class. Students can then work in pairs to complete Exercises 1–6. Make sure students understand that cosecant, secant, and cotangent represent the reciprocals of sine, cosine, and tangent, respectively. To review the answers to Exercises 3–6, you may wish to select students to write their verification on the chalkboard.

Answers
See Teacher Wraparound Edition page 391.

NAME _____ DATE _____ PERIOD ____

Mini-Project
(Use with Lesson 7-2.)

The Pythagorean Theorem

Trace the figure shown below and cut out the six parts. Then answer each of the following.

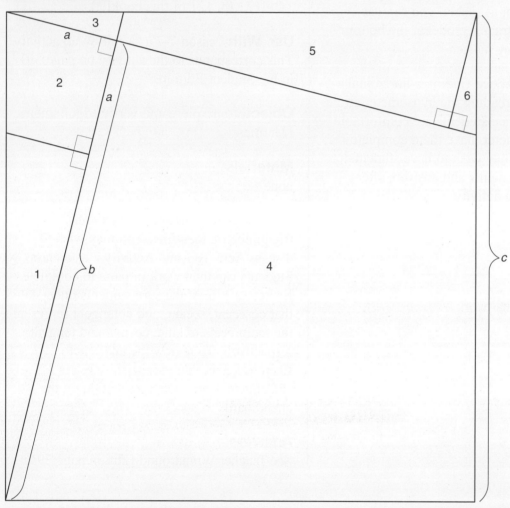

1. What is the area of the large square?

2. Show how two smaller squares can be made using all of the six parts.

3. What is the area of each of the smaller squares.

4. What is the relationship between the three squares.

5. Does $a^2 + b^2 = c^2$?

© Glencoe/McGraw-Hill   114   Teaching Geometry with Manipulatives

NAME _____ DATE _____ PERIOD _____

Geometry Activity Recording Sheet

(Use with the Lesson 7-2 Preview Activity on page 349 in the Student Edition.)

The Pythagorean Theorem

Materials
patty paper or square pieces of tracing paper
ruler

Model

1. Use a ruler to find actual measures for *a, b,* and *c*. Do these measures confirm that $a^2 + b^2 = c^2$?

2. Repeat the activity with different *a* and *b* values. What do you notice?

Analyze the Model

3. Explain why the drawing at the right is an illustration of the Pythagorean Theorem.

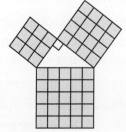

Using Overhead Manipulatives
(Use with Lesson 7-2.)

The Pythagorean Theorem

Objective Find the relationship among the sides of a right triangle.

Materials
- centimeter grids transparency*
- straightedge
- transparency pens*
- scissors*
- 2 blank transparencies

* = available in Overhead Manipulative Resources

Demonstration
Find a Triangle Relationship
- Place a blank transparency on the centimeter grid transparency. Use the straightedge to draw a segment 8 centimeters long. At one end of this segment, draw a perpendicular segment 6 centimeters long. Draw a third segment to form a triangle.
- Measure the length of the longest side of the triangle using the centimeter grid transparency. Ask students to state the length of this segment. **10 cm**
- On a blank transparency, draw three squares: one with 6 centimeters on a side, one with 8 centimeters on a side, and one with 10 centimeters on a side. Cut out the squares.
- Place the edges of each square against the corresponding side of the triangle.
- Ask students what kind of triangle was formed. **right triangle**
- Have students find the area of each square. **36 units^2, 64 units^2, 100 units^2** Write these areas on the squares.
- Ask students what relationship exists among the areas of the three squares. **36 + 64 = 100**

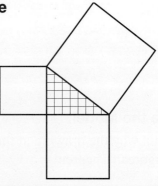

Extension
Test More Triangles
- Have students repeat the activity for each of the following triangles, and have them summarize their findings.
    - right triangle with perpendicular sides 9 units and 12 units long
    - nonright acute triangle with sides 9 units and 12 units long
    - obtuse triangle with sides 9 units and 12 units long

The area of the square along the hypotenuse of the triangle is equal to the sum of the areas of the squares on the other two sides. The area of the square on the longest side of the acute triangle is less than the sum of the areas of the other two squares. The area of the square on the longest side of the obtuse triangle is greater than the sum of the areas of the other two squares.

© Glencoe/McGraw-Hill   Teaching Geometry with Manipulatives

NAME _____ DATE _____ PERIOD ____

Geometry Activity Recording Sheet

(Use with the activity in Lesson 7-4 on page 365 in the Student Edition.)

Trigonometric Ratios

Materials
lined paper
scissors
centimeter ruler

Record the measures of each segment in the table.

Side	$\overline{AC}$	$\overline{AB}$	$\overline{BC}$	$\overline{AF}$	$\overline{AG}$	$\overline{AD}$	$\overline{AE}$	$\overline{DE}$
Measure								

Analyze

1. What is true of $\triangle AED$, $\triangle AGF$, and $\triangle ABC$?

2. Write the ratio of the side lengths for each trigonometric ratio in the table. Then calculate a value for each ratio to the nearest ten-thousandth.

	In $\triangle AED$	In $\triangle AGF$	In $\triangle ABC$
sin A			
cos A			
tan A			

3. Study the table. Write a sentence about the patterns you observe with the trigonometric ratios.

4. What is true about $m\angle A$ in each triangle?

NAME _____ DATE _____ PERIOD ____

Geometry Activity Transparency Master
(Use with Lesson 7-4.)

Trigonometry

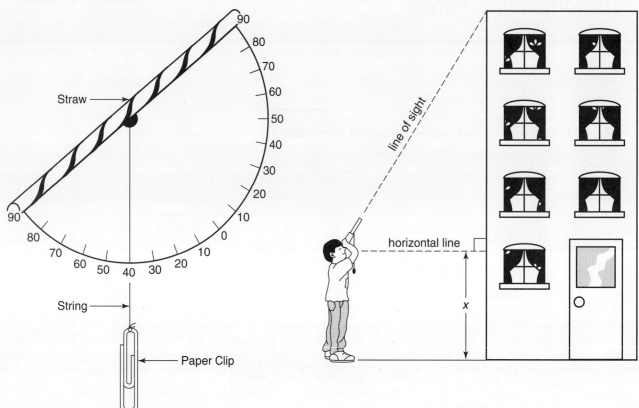

Place on the edge of the index card.

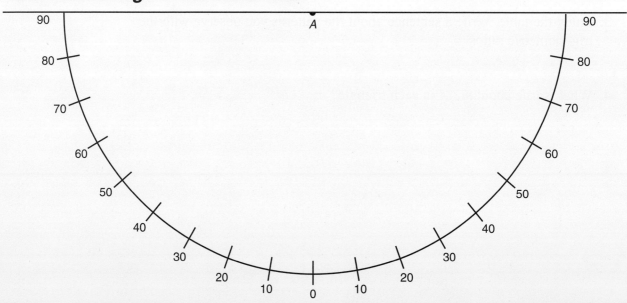

© Glencoe/McGraw-Hill

NAME _____ DATE _____ PERIOD ____

Geometry Activity

(Use with Lesson 7-4.)

Trigonometry

Find the height of each object. Use the calculator to find the approximate value of the tangent of the angle.

Example:

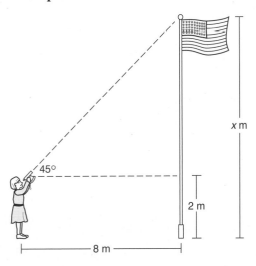

$\tan(45°) = \dfrac{x-2}{8}$

$1.0 = \dfrac{x-2}{8}$

$8 = x - 2$

$10 = x$

The flagpole is 10 meters high.

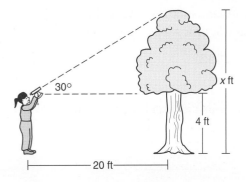

1. _____

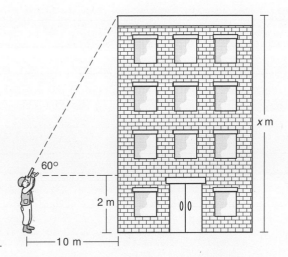

2. _____

© Glencoe/McGraw-Hill   Teaching Geometry with Manipulatives

Geometry Activity

Find the height of each of the following by using a calculator to find the approximate value of the tangent of the angle.

3. your school building _____

4. the flagpole _____

5. a tree _____

6. a telephone or an electric pole _____

7. the bleachers of the football field _____

8. another object

   (name of object) _____ (height of object) _____

9. sporting equipment, such as the field goal post on the football field, a basketball hoop, or a soccer net

   (name of equipment) _____ (height) _____

10. When would *you* use indirect measurement? _____

_____

_____

_____

_____

_____

_____

_____

_____

NAME _____ DATE _____ PERIOD _____

Geometry Activity Recording Sheet

(Use with the Lesson 7-7 Follow-Up Activity on page 391 in the Student Edition.)

Trigonometric Identities

Materials
none

Analyze

1. The identity $\cos^2 \theta + \sin^2 \theta = 1$ is known as a Pythagorean identity. Why do you think the word *Pythagorean* is used to name this?

2. Find two more reciprocal identities involving $\dfrac{1}{\cos \theta}$ and $\dfrac{1}{\tan \theta}$.

Verify each identity.

3. $\dfrac{\sin \theta}{\cos \theta} = \tan \theta$

4. $\cot \theta = \dfrac{\cos \theta}{\tan \theta}$

5. $\tan^2 \theta + 1 = \sec^2 \theta$

6. $\cot^2 \theta + 1 = \csc^2 \theta$

Chapter 8 Quadrilaterals
Teaching Notes and Overview

Using Overhead Manipulatives

Investigating the Exterior Angles of a Convex Polygon
(p. 127 of this booklet)

Use With Lesson 8-1.

Objective Investigate the sum of the measures of the exterior angles of a convex polygon.

Materials
straightedge
protractor*
transparency pens*
regular polygons transparency*
blank transparency
* = available in Overhead Manipulative Resources

This demonstration involves finding the sum of the measures of the exterior angles of a convex polygon. Students begin by making a conjecture that the sum of the measures of the exterior angles is 360 for regular polygons. As students examine more polygons at their desks, they may alter their conjecture to include any convex polygon.

In this extension, students work with a partner to prove their conjecture. You may wish to ask a few students to share their proofs with the class.

Answers
Answers appear on the teacher demonstration instructions on page 127.

Geometry Activity Recording Sheet

Sum of the Exterior Angles of a Polygon
(p. 128 of this booklet)

Use With the activity on page 406 in Lesson 8-1 of the Student Edition.

Objective Investigate the sum of the measures of the exterior angles of a convex polygon.

Materials
straightedge
protractor*
* = available in Overhead Manipulative Resources

For this activity, students draw five different convex polygons. You may wish to draw each type of polygon on the overhead to refer to while discussing the table in Exercise 1. This activity should lead students to the conclusion stated in Theorem 8.2.

Answers
See Teacher Wraparound Edition page 406.

Geometry Activity Recording Sheet

Properties of Parallelograms
(p. 129 of this booklet)

Use With the activity on pages 411–412 in Lesson 8-2 of the Student Edition.

Objective Use patty paper to examine properties of parallelograms.

Materials
patty paper or tracing paper
straightedge

For this activity, students can work in pairs. Students construct a parallelogram on one piece of patty paper, and then trace the parallelogram on another piece of patty paper. Students identify congruent angles and congruent segments of the two parallelograms, and come to the conclusion that opposite angles and opposite sides are congruent.

Answers
See Teacher Wraparound Edition page 412.

Chapter 8 Teaching Notes and Overview

Using Overhead Manipulatives

Tests for Parallelograms
(pp. 130–131 of this booklet)

Use With Lesson 8-3.

Objective Investigate conditions that ensure that a quadrilateral is a parallelogram.

Materials
pipe cleaners or twist ties
lined paper
geoboard*
geobands*
protractor*
blank transparency
* = available in Overhead Manipulative Resources

This activity includes two demonstrations that test whether a quadrilateral is a parallelogram.
- Demonstration 1 uses pipe cleaners or twist ties to construct a quadrilateral with opposite sides congruent. By measuring the angles of the quadrilateral, students discover that adjacent angles have a sum of 180, and therefore, opposite sides are parallel. You may wish to pass out pipe cleaners or twist ties to students and have them complete the activity at their desks.
- Demonstration 2 uses a geoboard and geobands to form parallelograms. You may wish to have students come to the overhead and manipulate the geobands to form different parallelograms.

Answers
Answers appear on the teacher demonstration instructions on pages 130–131.

Geometry Activity Recording Sheet

Testing for a Parallelogram
(p. 132 of this booklet)

Use With the activity on page 417 in Lesson 8-3 of the Student Edition.

Objective Create a quadrilateral model and examine properties that determine whether it is a parallelogram.

Materials
straws
pipe cleaners
ruler*
protractor*
* = available in Overhead Manipulative Resources

For this activity, students use straws and pipe cleaners to create a quadrilateral that can be shifted to form quadrilaterals of different shapes. Students measure distances and angles, which allow them to classify the quadrilateral as a parallelogram. Discuss with students the properties that are necessary to determine that a quadrilateral is a parallelogram.

Answers
See Teacher Wraparound Edition page 417.

Using Overhead Manipulatives

Constructing a Rectangle
(p. 133 of this booklet)

Use With Lesson 8-4.

Objective Construct a rectangle.

Materials
compass*
straightedge
transparency pens*
blank transparency
* = available in Overhead Manipulative Resources

© Glencoe/McGraw-Hill 124 Teaching Geometry with Manipulatives

Chapter 8 Teaching Notes and Overview

This demonstration involves using a compass and straightedge to construct a rectangle. You may wish to have students complete the construction at their desks while you complete the construction at the overhead. Discuss with students why the figure is both a rectangle and a parallelogram.

Answers
Answers appear on the teacher demonstration instructions on page 133.

Using Overhead Manipulatives
Constructing a Rhombus
(p. 134 of this booklet)

Use With Lesson 8-5.

Objective Construct a rhombus.

Materials
compass*
straightedge
transparency pens*
blank transparency
* = available in Overhead Manipulative Resources

This demonstration involves using a compass and straightedge to construct a rhombus. You may wish to have students complete the construction at their desks while you complete the construction at the overhead. Discuss with students how they can use the method for constructing a rectangle and constructing a rhombus to construct a square.

Answers
Answers appear on the teacher demonstration instructions on page 134.

Mini-Project
Square Search
(p. 135 of this booklet)

Use With Lesson 8-5.

Objective Determine the total number of squares in each figure.

Have students work in groups of two or three to count the number of squares in each figure. Tell students that it may be helpful to use colored pencils to trace each square as you count them. This may help prevent counting the same square more than once.

Answers
1. 55
2. 8
3. 22
4. 29

Geometry Activity Recording Sheet
Kites
(p. 136 of this booklet)

Use With Lesson 8-5 as a follow-up activity. This corresponds to the activity on page 438 in the Student Edition.

Objective Explore properties of kites.

Materials
compass*
ruler*
protractor*
* = available in Overhead Manipulative Resources

In this activity, students use a compass and straightedge to construct a kite and its diagonals. By measuring segments and angles,

Chapter 8 Teaching Notes and Overview

students discover properties of kites. Once students have completed the activity, ask volunteers to list one of their conjectures about the properties of kites on the chalkboard or overhead.

Answers
See Teacher Wraparound Edition page 438.

Geometry Activity Recording Sheet
Construct Median of a Trapezoid
(p. 137 of this booklet)

Use With the activity on page 441 in Lesson 8-6 of the Student Edition.

Objective Construct the median of a trapezoid.

Materials
compass*
ruler*
* = available in Overhead Manipulative Resources

Students begin this activity by drawing a trapezoid and constructing its median. By measuring the bases and median, students are led to the conclusion that the measure of the median of a trapezoid is equal to half the sum of the measures of the bases. You may wish to select a student to perform the construction on the overhead, and then select another student to measure the lengths of the three segments.

Answers
See Teacher Wraparound Edition page 441.

Geometry Activity
Linear Equations
(pp. 138–139 of this booklet)

Use With Lesson 8-7.

Objective Identify the quadrilateral formed by the midpoints of a quadrilateral as a parallelogram.

Materials
classroom set of Geometry Activity worksheets
unlined paper
ruler*
scissors*
* = available in Overhead Manipulative Resources

In this activity, students use two different models to make a conjecture about the quadrilateral formed by the midpoints of a quadrilateral. For the first model, students cut out and fold the midpoints of each side of a quadrilateral. By measuring the lengths of the sides of the new quadrilateral formed by the folds, students discover that opposite sides are congruent. In the second model, students graph a quadrilateral on a coordinate grid. Then by using the Midpoint Formula and the Distance Formula, students discover again that opposite sides are congruent. Through both models and Theorem 8.9, students can make a conjecture that a parallelogram is formed by the midpoints of a quadrilateral.

Answers
1–5. See students' work.

6. Parallelogram; if both pairs of opposite sides of a quadrilateral are congruent, then the quadrilateral is a parallelogram.

7–12. See students' work.

13. Parallelogram; if both pairs of opposite sides of a quadrilateral are congruent, then the quadrilateral is a parallelogram.

14. A parallelogram is formed by the midpoints of a quadrilateral.

Using Overhead Manipulatives
(Use with Lesson 8-1.)

Investigating the Exterior Angles of a Convex Polygon

Objective Investigate the sum of the measures of the exterior angles of a convex polygon.

Materials
- straightedge
- protractor*
- transparency pens*
- regular polygons transparency*
- blank transparencies

* = available in Overhead Manipulative Resources

Demonstration
Investigate the Exterior Angles of a Convex Polygon
- Place a blank transparency over the regular polygons transparency and trace the octagon.

- Using a different colored transparency pen, extend the sides of the octagon to form one exterior angle at each vertex.

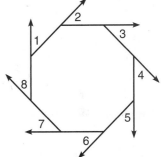

- Have students use a protractor to find the measure of each exterior angle. Then have them find the sum of the measures and record on the transparency.

- Repeat the process on a regular decagon and on a regular dodecagon.

- Ask students to make a conjecture about the sum of the measures of the exterior angles of a regular polygon. **The sum of the measures of the exterior angles is 360.**

- Have students work in pairs to draw a few more regular polygons to test their conjecture. Suggest that they try a nonregular polygon. Ask students if the result is the same. **yes**

Extension
Prove the Conjecture
- Have students work in pairs to prove their conjecture. Suggest that they consider the linear pairs of angles formed by the interior angles of the polygon and their corresponding exterior angles.

© Glencoe/McGraw-Hill   Teaching Geometry with Manipulatives

NAME _____ DATE _____ PERIOD ____

Geometry Activity Recording Sheet

(Use with the activity on page 406 in Lesson 8-1 of the Student Edition.)

Sum of the Exterior Angles of a Polygon

Materials
straightedge
protractor

Collect Data
Draw each polygon in the space provided.
triangle   convex quadrilateral   convex pentagon

convex hexagon   convex heptagon

Analyze the Data

1. Complete the table.

Polygon	triangle	quadrilateral	pentagon	hexagon	heptagon
number of exterior angles					
sum of measures of exterior angles					

2. What conjecture can you make?

© Glencoe/McGraw-Hill   128   Teaching Geometry with Manipulatives

NAME _____ DATE _____ PERIOD ____

Geometry Activity Recording Sheet

(Use with the activity on pages 411–412 in Lesson 8-2 of the Student Edition.)

Properties of Parallelograms

Materials
patty paper or tracing paper
straightedge

Analyze the Model

1. List all of the segments that are congruent.

2. List all of the angles that are congruent.

3. Describe the angle relationships you observe.

© Glencoe/McGraw-Hill   Teaching Geometry with Manipulatives

Using Overhead Manipulatives
(Use with Lesson 8-3.)

Tests for Parallelograms

Objective Investigate some of the conditions that ensure that a quadrilateral is a parallelogram.

Materials
- pipe cleaners or twist ties
- lined paper
- geoboard*
- geobands*
- protractor*
- blank transparency

* = available in Overhead Manipulative Resources

Demonstration 1
Use Pipe Cleaners to Ensure a Quadrilateral is a Parallelogram

- Cut two pairs of congruent pipe cleaners and twist the ends together to form a quadrilateral with opposite sides congruent. Place on a blank transparency and label the quadrilateral *ABCD*. Point out that both pairs of opposite sides are congruent.

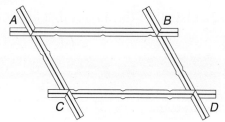

- Use the protractor to measure each angle in the quadrilateral. Record the angle measures on the transparency. (Sample measures are given.)

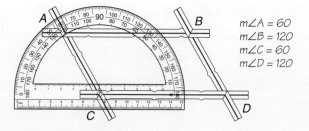

$m\angle A = 60$
$m\angle B = 120$
$m\angle C = 60$
$m\angle D = 120$

- Ask students what they notice about the measures of angles *A* and *C* and angles *B* and *D*. **They have a sum of 180.** What can they conclude about $\overline{AB}$ and $\overline{CD}$ and $\overline{AC}$ and $\overline{BD}$ as a result? **$\overline{AB} \parallel \overline{CD}$ and $\overline{AB} \parallel \overline{BD}$ since when cut by a transversal their consecutive interior angles are supplementary.** Therefore, the quadrilateral is a parallelogram by definition.
- Simultaneously pull on opposite vertices of the quadrilateral to change its shape. Then repeat the steps listed above. Is the result the same? **yes**

© Glencoe/McGraw-Hill   Teaching Geometry with Manipulatives

Using Overhead Manipulatives

Demonstration 2
Use a Geoboard to Ensure a Quadrilateral is a Parallelogram
- Place the geoboard on the overhead. Place a geoband around pegs $L(2, 3)$ and $I(4, 4)$ using the numbers along the edges of the geoboard.

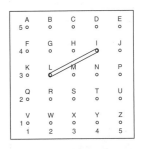

- Leaving the geoband around L and I, grasp the geoband and ask a student to tell you how many units to move up or down from L and I to form a parallelogram. Move the geoband accordingly.

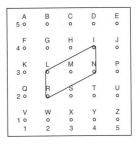

- Still holding the geoband, ask a second student to tell you how many units to move left or right from L and I. Move the geoband accordingly, and place the geoband around the points found. This will form a different parallelogram.

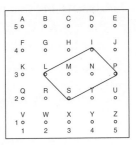

- The third and fourth points were found in such a way that segments between these points and L and I must be congruent and parallel. Why? **The number of units moved up or down and left or right determines the same slope and measure.** Ask students to verify that the slopes and measure of these segments are equal. Point out that this means that this quadrilateral meets the conditions of the theorem that says that a quadrilateral with two opposite sides parallel and congruent is a parallelogram.
- Ask two more students to find the slopes of $\overline{LI}$ and the segment between the third and fourth points. The slopes are equal, so these segments are parallel. Thus, the quadrilateral is a parallelogram.
- You may wish to repeat the investigation with four student-generated points.

© Glencoe/McGraw-Hill   131   Teaching Geometry with Manipulatives

NAME _____ DATE _____ PERIOD _____

Geometry Activity Recording Sheet

(Use with the activity on page 417 in Lesson 8-3 of the Student Edition.)

Testing for a Parallelogram

Materials
straws
pipe cleaners
ruler
protractor

Analyze

1. Measure the distance between the opposite sides of the quadrilateral in at least three places. Repeat this process for several figures. What can you conclude about opposite sides?

2. Classify the quadrilaterals that you formed.

3. Compare the measures of pairs of opposite sides.

4. Measure the four angles in several of the quadrilaterals. What relationships do you find?

Make a Conjecture

5. What conditions are necessary to verify that a quadrilateral is a parallelogram?

Using Overhead Manipulatives
(Use with Lesson 8-4.)

Constructing a Rectangle

Objective Construct a rectangle.

Materials
- compass*
- straightedge
- transparency pens*
- blank transparency

* = available in Overhead Manipulative Resources

Demonstration
Construct a Rectangle
- Use a straightedge to draw a line a. Choose a point D on a.
- With the compass at 8 centimeters, place the point at D and locate point E on a so that $DE = 8$ centimeters. Construct lines perpendicular to a through D and E. You may wish to refer students to the construction of a perpendicular line through a point in Lesson 1-5 of *Glencoe Geometry*. Label the lines b and c.

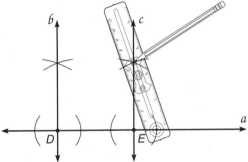

- Set the compass to 5 centimeters. Place the compass point at D and mark off a segment on b. Using the same compass setting, place the compass point at E and mark a segment on c. Label these points F and G, respectively.

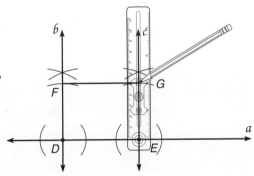

- Use a straightedge to draw $\overline{FG}$.
- Ask students to explain why $DEGF$ is a parallelogram. **$\overline{DF} \parallel \overline{EG}$ since each is perpendicular to a, and if two lines are perpendicular to the same line, then they are parallel. $\overline{DF} \cong \overline{EG}$ by construction. Therefore, $DEGF$ is a parallelogram because two opposite sides are parallel and congruent.**
- Ask the students to explain why $DEGF$ is a rectangle. **Quadrilateral $DEGF$ is a parallelogram and all angles are right angles.**

Using Overhead Manipulatives
(Use with Lesson 8-5.)

Constructing a Rhombus

Objective Construct a rhombus.

Materials
- compass*
- straightedge
- transparency pens*
- blank transparency

* = available in Overhead Manipulative Resources

Demonstration
Construct a Rhombus

- Draw $\overline{WZ}$. Set the compass to match the length of $\overline{WZ}$. Use this compass setting for all arcs drawn.

- Place the compass at point W and draw an arc above $\overline{WZ}$. Choose any point on the arc and label it X.

- Place the compass at point X and draw an arc to the right of X. Then place the compass at point Z and draw an arc to intersect the arc drawn from point X. Label the point of intersection Y.

- Use a straightedge to draw $\overline{WX}$, $\overline{XY}$, and $\overline{YZ}$.

- Ask students to explain why $WXYZ$ is a rhombus.
  All four sides are congruent.

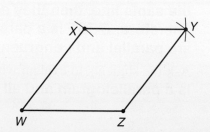

© Glencoe/McGraw-Hill   134   Teaching Geometry with Manipulatives

NAME _____ DATE _____ PERIOD ____

Mini-Project
(Use with Lesson 8-5.)

Square Search

Work together to determine the total number of squares in each figure.

1.

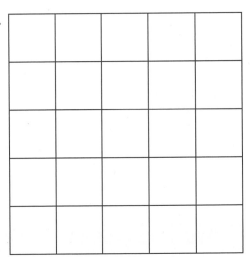

2.

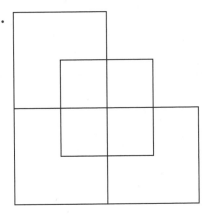

3.

4.

© Glencoe/McGraw-Hill   135   Teaching Geometry with Manipulatives

NAME _____ DATE _____ PERIOD ____

Geometry Activity Recording Sheet

(Use with the Lesson 8-5 Follow-Up Activity on page 438 in the Student Edition.)

Kites

Materials
compass
ruler
protractor

Activity
Construct a kite *QRST* in the space provided at the right.

Model

1. Draw $\overline{QS}$ in kite *QRST*. Use a protractor to measure the angles formed by the intersection of $\overline{QS}$ and $\overline{RT}$.

2. Measure the interior angles of kite *QRST*. Are any congruent?

3. Label the intersection of $\overline{QS}$ and $\overline{RT}$ as point *N*. Find the lengths of $\overline{QN}$, $\overline{NS}$, $\overline{TN}$, and $\overline{NR}$. How are they related?

4. How many pairs of congruent triangles can be found in kite *QRST*?

5. Construct another kite *JKLM*. Repeat Exercises 1-4.

Analyze

6. Use your observations and measurements of kites *QRST* and *JKLM* to make conjectures about the angles, sides, and diagonals of kites.

© Glencoe/McGraw-Hill   Teaching Geometry with Manipulatives

NAME _____ DATE _____ PERIOD _____

Geometry Activity Recording Sheet

(Use with the activity on page 441 in Lesson 8-6 of the Student Edition.)

Construct Median of a Trapezoid

Materials
compass
ruler

Model
Draw trapezoid WXYZ, and construct median MN in the space provided below.

Analyze

1. Measure $\overline{WX}$, $\overline{ZY}$, and $\overline{MN}$ to the nearest millimeter. Record your measures in the table.

Side	Measure
$\overline{WX}$	
$\overline{ZY}$	
$\overline{MN}$	

2. **Make a conjecture** based on your observations.

NAME _____ DATE _____ PERIOD ____

Geometry Activity

(Use with Lesson 8-7.)

Linear Equations

1. On a separate piece of paper, draw quadrilateral *ABCD* with no two sides parallel or congruent. Cut out quadrilateral *ABCD*.

2. Fold to find the midpoint of $\overline{AB}$ and pinch to make a crease at the point. Mark the point with a pencil.

3. Fold to find the midpoints of $\overline{BC}$, $\overline{CD}$, and $\overline{DA}$. Mark the points.

4. Draw the quadrilateral formed by the midpoints of the segments.

5. Measure each side of the quadrilateral determined by the midpoints. Record your measurements on the sides.

6. What type of quadrilateral do the folds in quadrilateral *ABCD* form? Justify your answer.

7. Graph a quadrilateral *WXYZ* with no two sides parallel or congruent on the coordinate grid.

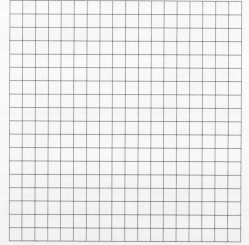

© Glencoe/McGraw-Hill   138   Teaching Geometry with Manipulatives

Geometry Activity

8. Use the Midpoint Formula to find the midpoint of $\overline{WX}$. Label the midpoint P.

9. Find the midpoints of $\overline{XY}$, $\overline{YZ}$, and $\overline{ZW}$. Label the midpoints Q, R, and S, respectively. Record the coordinates of the midpoints in the table.

Midpoint	Coordinates
P	
Q	
R	
S	

10. Draw $\overline{PQ}$, $\overline{QR}$, $\overline{RS}$, and $\overline{SP}$ to form a quadrilateral within $WXYZ$.

11. Use the Distance Formula to find the distance of $\overline{PQ}$.

12. Find the distance of $\overline{QR}$, $\overline{RS}$, and $\overline{SP}$. Record the distances in the table.

Side	Measure
$\overline{PQ}$	
$\overline{QR}$	
$\overline{RS}$	
$\overline{SP}$	

13. What type of quadrilateral is $PQRS$? Justify your answer.

14. **Make a conjecture** about the type of quadrilateral formed by the midpoints of a quadrilateral.

Chapter 9
Transformations
Teaching Notes and Overview

Geometry Activity Recording Sheet
Transformations
(p. 145 of this booklet)

Use With Lesson 9-1 as a preview activity. This corresponds to the activity on page 462 in the Student Edition.

Objective Identify transformations as a translation, reflection, rotation, or dilation.

Materials
none

For this activity, students examine translations and identify them by the specific type. Remind students that it is possible that each illustration could represent more than one type of transformation. Students also determine that rotations, reflections, and translations represent isometries.

Answers
See Teacher Wraparound Edition page 462.

Using Overhead Manipulatives
Constructing Reflections in a Line
(p. 146 of this booklet)

Use With Lesson 9-1.

Objective Create a reflection using dot paper.

Materials
dot paper transparency*
transparency pens*
straightedge
dot paper for students
* = available in Overhead Manipulative Resources

This demonstration involves using dot paper to create a reflection image. After you plot J', you may wish to select students to come to the overhead and plot K', L', and M'. In the extension, quadrilaterals $JKLM$ and $J'K'L'M'$ are reflected in a horizontal line. Students then work in pairs to complete reflections of images drawn by their partner on dot paper.

Answers
Answers appear on the teacher demonstration instructions on page 146.

Using Overhead Manipulatives
Translations
(pp. 147–148 of this booklet)

Use With Lesson 9-2.

Objective Use translation images to draw a prism.

Materials
lined paper transparency*
regular polygons transparency*
coordinate grids transparency*
transparency pens*
straightedge
blank transparency
* = available in Overhead Manipulative Resources

This activity involves two demonstrations to draw a prism.
- Demonstration 1 uses the lined paper transparency and regular polygons transparency to draw a pentagonal prism. You may wish to have students follow along at their desks on a piece of lined paper. Ask students to explain how the pentagon was translated.
- Demonstration 2 uses the coordinate grids transparency to draw a triangular prism. Again, you may choose for students to

Chapter 9 Teaching Notes and Overview

follow along at their desks on a piece of grid paper.
- In this extension, students work in pairs to practice drawing translation images described by an ordered pair.

Answers
Answers appear on the teacher demonstration instructions on pages 147–148.

 Geometry Activity
Reflections and Translations
(pp. 149–150 of this booklet)

Use With Lesson 9-2.

Objective Use a geomirror to discover that two successive reflections in a pair of parallel lines is the same as a translation from the preimage.

Materials
classroom set of Geometry Activity worksheets
unlined paper
geomirror*
straightedge
* = available in Overhead Manipulative Resources

For this activity, students can work in groups of two or three. Before giving students the worksheets, you may wish to pass out and review how to use the geomirrors.

In this activity, students reflect and draw a figure in each of a pair of parallel lines. By examining the preimage and images, students make a conjecture that the resulting image from the pair of reflections is the same as translating the preimage.

Answers
1–5. See students' work.

6. The orientation remains unchanged.
7. yes
8. $AA'' = BB'' = CC'' = 2$(distance from ℓ to m)
9. Use a translation.
10. See students' work.
11. See students' work.
12. yes

 Mini-Project
Graphing and Translations
(p. 151 of this booklet)

Use With Lesson 9-2.

Objective Perform translations and reflections of triangles and line segments.

For this activity, students should work in pairs. For Exercises 1 and 2, students graph triangles and their translated images. For Exercises 3 and 4, students draw line segments and their reflected images. Students should discover the relationship of the slope of a line segment and its image after a reflection in the y-axis and then in the x-axis.

Answers
1.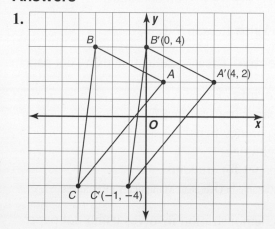

2.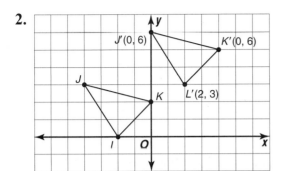

3.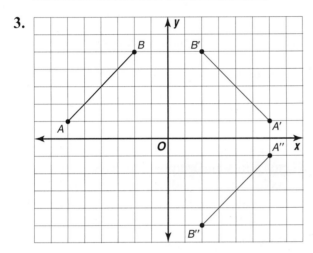

(6, −1), (2, −5)

4.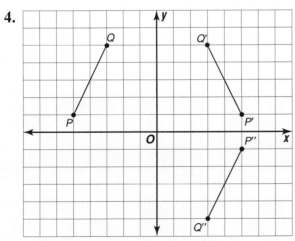

Slopes are $\overline{PQ}$: 2, $\overline{P'Q'}$: −2, $\overline{P''Q''}$: 2; the slopes of $\overline{PQ}$ and $\overline{P''Q''}$ are equal.

 Using Overhead Manipulatives

Rotations
(pp. 152–153 of this booklet)

Use With Lesson 9-3.

Objective Find rotation images.

Materials
straightedge
transparency pens*
blank transparencies

* = available in Overhead Manipulative Resources

This demonstration involves reflecting an image in a pair of intersecting lines to discover that this is the same as a rotation about the point of intersection of the two lines. You may wish to distribute patty paper or wax paper and have students complete the activity at their desks while you complete it at the overhead.

In this extension, students investigate the angle of rotation when an image is reflected in perpendicular lines.

Answers
Answers appear on the teacher demonstration instructions on pages 152–153.

 Geometry Activity Recording Sheet

Tessellations of Regular Polygons
(p. 154 of this booklet)

Use With the activity on page 483 in Lesson 9-4 of the Student Edition.

Objective Study pattern blocks to determine which regular polygons tessellate.

Chapter 9 Teaching Notes and Overview

Materials
pattern blocks*

* = available in Overhead Manipulative Resources

For this activity, students can work in groups of three to four. Students use polygons of one type from the set of pattern blocks to try and form tessellations. Once students complete the activity, have a volunteer from each group share the groups conjecture with the rest of the class.

Answers
See Teacher Wraparound Edition page 483.

Geometry Activity Recording Sheet

Tessellations and Transformations
(p. 155 of this booklet)

Use With Lesson 9-4 as a follow-up activity. This corresponds to the activity on page 489 in the Student Edition.

Objective Use transformations to create tessellations.

Materials
unlined paper

In this activity, students are given directions to create two different tessellations. The first tessellation is created by translating an image. The second tessellation is created by rotating the image. You may encourage students to use different colors in each tessellation. Ask for volunteers to show their tessellations to the class.

Answers
See Teacher Wraparound Edition page 489.

Geometry Activity Recording Sheet

Comparing Magnitude and Components of Vectors
(p. 156 of this booklet)

Use With the activity on page 501 in Lesson 9-6 of the Student Edition.

Objective Determine how changing the magnitude affects the component of a vector.

Materials
grid paper
straightedge

Students begin this activity by drawing a vector $\vec{a}$ in standard position. Then they draw a vector with twice the magnitude of $\vec{a}$ and a vector with three times the magnitude of $\vec{a}$. By examining the components of these three vectors, students discover how multiplying by a scalar changes the magnitude of a vector.

Answers
See Teacher Wraparound Edition page 501.

NAME _____ DATE _____ PERIOD ____

Geometry Activity Recording Sheet

(Use with the Lesson 9-1 Preview Activity on page 462 in the Student Edition.)

Transformations

Materials
none

Exercises
Identify each transformation. Refer to page 462 in the Student Edition to examine each transformation.

1.

2.

3.

4.

5.

6.

7.

8.

9.

10.

Make a Conjecture

11. An isometry is a transformation in which the resulting image is congruent to the preimage. Which transformations are isometries?

Using Overhead Manipulatives
(Use with Lesson 9-1.)

Constructing Reflections in a Line

Objective Create a reflection using dot paper.

Materials
- dot paper transparency*
- transparency pens*
- straightedge
- dot paper for students

* = available in Overhead Manipulative Resources

Demonstration
Create a Reflection in a Vertical Line

- Draw the quadrilateral shown on the dot paper transparency. Use a colored transparency pen to draw a vertical line of reflection.

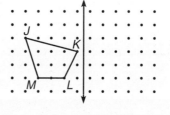

- Tell students that vertex *J* is 5 dots to the left of the line of reflection. Have students find a dot on the same row that is 5 dots to the right of the line of reflection. Label the dot *J'*.

- Repeat the steps for vertices *K, L,* and *M.* Draw lines between the vertices to complete the reflection.

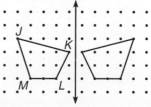

- Have students compare the two figures. Ask them if the corresponding sides are congruent. **yes**

- Ask, "How is this reflection like a reflection in a mirror?" **Each point is the same distance from the line of reflection as its corresponding point. In a mirror, your reflection seems to be as far "into" the mirror as you are in front of the mirror.**

Extension
Create a Reflection in a Horizontal Line

- Add a horizontal line of reflection below the figures on the dot paper transparency. Reflect *JKLM* and *J'K'L'M'* in the horizontal line of reflection.

- Draw a triangle and a line of reflection that is neither vertical nor horizontal on the dot paper transparency. Have students draw the figure on dot paper. Students should work in pairs to complete the reflection.

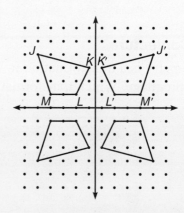

© Glencoe/McGraw-Hill   146   Teaching Geometry with Manipulatives

Using Overhead Manipulatives
(Use with Lesson 9-2.)

Translations

Objective Use translation images to draw a prism.

Materials
- lined paper transparency*
- regular polygons transparency*
- coordinate grids transparency*
- transparency pens*
- straightedge
- blank transparency

* = available in Overhead Manipulative Resources

Demonstration 1
Draw a Prism Using a Translation Image
- Place the lined paper transparency over the regular polygons transparency and align the bottom of the pentagon with a line on the lined paper transparency. Trace the pentagon.

- Remove the regular polygons transparency. Lay a blank transparency over the lined paper transparency and trace the pentagon again.
- Slowly slide the blank transparency toward the upper right corner of the lined paper transparency. Stop when the bottom of the image pentagon is aligned with a different line of the lined paper.
- Use a straightedge to connect the corresponding vertices of the two pentagons. Make the segments representing sides of the solid that will be hidden from view dashed.

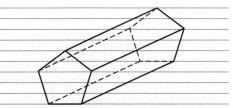

Demonstration 2
Draw a Prism Using a Translation Image
- Copy the graph below on the coordinate grids transparency.

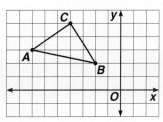

Using Overhead Manipulatives
- Lay a blank transparency over the coordinate grids transparency and copy the lines of △ABC onto the transparency with a transparency pen.

- Translate the image triangle 4 units to the left. Then translate the image 3 units down. Label the vertices of the image triangle A′, B′, and C′.

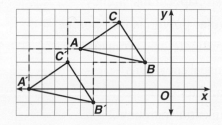

- Ask students to name the coordinates of the vertices of the image triangle.
  A′(−11, 0), B′(−6, −1), C′(−8, 2)

- Ask students to find the slope of the lines between A, B, and C and the corresponding vertices in the image triangle. **The slopes are all $\frac{3}{4}$.**

Extension
Describe Translations as Ordered Pairs
- Tell students that it is possible to describe a translation as an ordered pair. The translation from the demonstration can be described as (−4, −3). Have students work in pairs. Each student should draw a preimage on a coordinate grid and give an ordered pair for a translation. Then exchange papers and find the images.

NAME _____ DATE _____ PERIOD _____

Geometry Activity

(Use with Lesson 9-2.)

Reflections and Translations

A geomirror is a construction instrument that allows you to find the reflection image of a figure. Use a geomirror with the following activity to investigate translations.

1. On a separate piece of paper, draw two parallel lines ℓ and m and a triangle ABC.

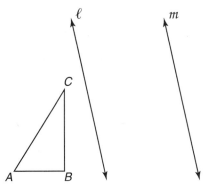

2. Place the geomirror so that the edge is aligned with line ℓ. Look into the geomirror to see the reflection image of $\triangle ABC$.

3. Use a straightedge to draw the image of $\triangle ABC$. Label the vertices A', B', and C'.

4. Align the edge of the geomirror with line m. Look into the geomirror to see the reflection image of $\triangle A'B'C'$.

5. Draw the image of $\triangle A'B'C'$. Label the vertices A'', B'', and C''.

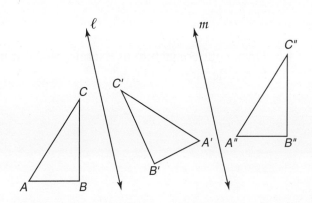

© Glencoe/McGraw-Hill   149   Teaching Geometry with Manipulatives

Geometry Activity

Refer to your drawing for Exercises 6–9.

6. You know that a reflection changes the orientation of its image. What happens when a figure is reflected twice?

7. Are the points A, A', and A' collinear? How about B, B', and B'' and C, C', and C'''? Use a straightedge to verify your answer.

8. Compare the lengths of $\overline{AA''}$, $\overline{BB''}$, and $\overline{CC''}$, and the distance between ℓ and m.

9. Describe how you could map $\triangle ABC$ onto $\triangle A''B''C''$ in one motion instead of two reflections.

Use the geomirror to test whether two reflections will translate a regular polygon.

10. Draw two parallel lines and a regular polygon.

11. Reflect the regular polygon twice and compare the preimage and the image.

12. Does your conjecture in Exercise 9 about mapping in one motion instead of two reflections hold true?

NAME _____ DATE _____ PERIOD ____

Mini-Project
(Use with Lesson 9-2.)

Graphing and Translations

Graph paper can help you draw transformation images of figures.

1. Graph △ABC with vertices A(1, 2), B(−3, 4), and C(−4, −4). Draw △A′B′C′, the translation image of △ABC where the distance of the slide is 3 units to the right. Name the coordinates of the translation image of each vertex.

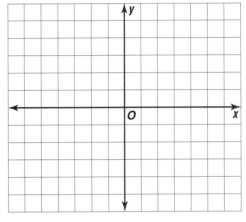

2. Draw △JKL with vertices J(−4, 3), K(0, 2), and L(−2, 0). Let △J′K′L′ be the image of △JKL under a slide of 4 units to the right followed by a slide of 3 units up. Graph △J′K′L′. Name the coordinates of the vertices of △J′K′L′.

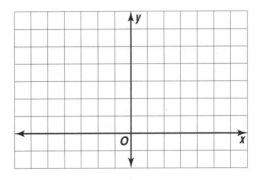

3. Draw $\overline{A'B'}$, the image formed by reflecting $\overline{AB}$ in the y-axis. Then draw $\overline{A''B''}$, the image formed by reflecting $\overline{A'B'}$ in the x-axis. What are the coordinates of A″ and B″?

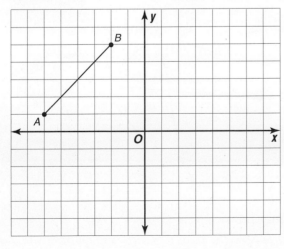

4. Draw $\overline{P'Q'}$, the reflection image of $\overline{PQ}$ in the y-axis. Draw $\overline{P''Q''}$, the reflection image of $\overline{P'Q'}$ in the x-axis. Find slopes of $\overline{PQ}$, $\overline{P'Q'}$, and $\overline{P''Q''}$. What is the relationship between the slopes of $\overline{PQ}$ and $\overline{P''Q''}$?

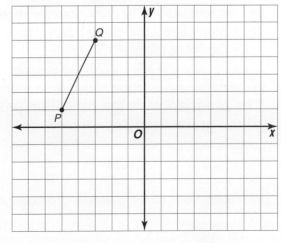

© Glencoe/McGraw-Hill   Teaching Geometry with Manipulatives

Using Overhead Manipulatives
(Use with Lesson 9-3.)

Rotations

Objective Find rotation images.

Materials
- straightedge
- transparency pens*
- blank transparencies

* = available in Overhead Manipulative Resources

Demonstration
Find Rotation Images about Intersecting Lines

- Use a blue transparency pen to draw a triangle and two intersecting lines ℓ and m on a blank transparency. Label the point of intersection of lines ℓ and m as point A.

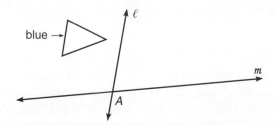

- Fold the transparency along line ℓ so that the side containing the triangle is on the bottom. Use a green transparency pen to trace the triangle. Unfold the transparency and ask students to describe the green triangle in terms of the blue triangle. **The green triangle is the reflection image of the blue triangle in line ℓ.**

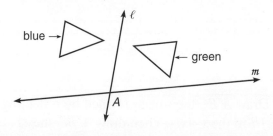

- Fold the transparency along line m so that the side containing the green triangle is on the bottom. Use a red transparency pen to trace the green triangle. Unfold the transparency and ask students to describe the red triangle in terms of the green triangle. **The red triangle is the reflection image of the green triangle in line m.**

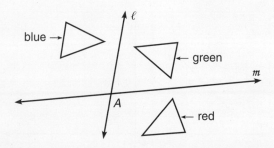

- Tell students that the red triangle is a composition of two reflections of the blue triangle in the intersecting lines.

© Glencoe/McGraw-Hill

Teaching Geometry with Manipulatives

Using Overhead Manipulatives

- Place a second blank transparency over the diagram. Trace the blue triangle with a black transparency pen. Placing your pen point at point *A,* rotate the top transparency about point *A* until the black triangle coincides with the red triangle on the bottom transparency. Ask students to make a conjecture about the composition of reflections in two intersecting lines. **The image of two reflections in intersecting lines is the same as a rotation about the point of intersection of the two lines of reflection.**

Extension
Find Rotation Images about Perpendicular Lines

- Have students draw a preimage and two perpendicular lines *s* and *t*. Have them reflect the preimage in *s* and then in *t*. Then place a piece of unlined or waxed paper over the figure and trace the preimage. Rotate the preimage until it coincides with the image of the two reflections. Ask, "How many degrees must the paper be turned so the preimage and image coincide?" **180** Ask, "How is this related to the measure of the angle between the two lines of reflections?" **It is twice the measure of the angle between the lines of reflection.**

NAME _____ DATE _____ PERIOD _____

Geometry Activity Recording Sheet

(Use with the activity on page 483 in Lesson 9-4 of the Student Edition.)

Tessellations of Regular Polygons

Materials
pattern blocks

Model and Analyze

1. Which shapes in the set are regular?

2. Write an expression showing the sum of the angles at each vertex of the tessellation.

3. Complete the table below.

Regular Polygon	triangle	square	pentagon	hexagon	heptagon	octagon
Measure of One Interior Angle						
Does it tessellate?						

Make a Conjecture

4. What must be true of the angle measure of a regular polygon for a regular tessellation to occur?

NAME _____ DATE _____ PERIOD ____

Geometry Activity Recording Sheet

(Use with the Lesson 9-4 Follow-Up Activity on page 489 in the Student Edition.)

Tessellations and Transformations

Materials
unlined paper

Model and Analyze

1. Is the area of the square in Step 1 of Activity 1 the same as the area of the new shape in Step 2? Explain.

2. Describe how you would create the unit for the pattern shown at the right.

Make a tessellation for each pattern described. Use a tessellation of two rows of three squares as your base.

3.

4.

5.

NAME _____ DATE _____ PERIOD ____

Geometry Activity

(Use with the activity on page 501 in Lesson 9-6 of the Student Edition.)

Comparing Magnitude and Components of Vectors

Materials
grid paper
straightedge

Model and Analyze
Draw $\vec{a}$ in standard position.

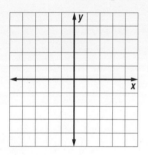

Draw $\vec{b}$ in standard position with the same direction as $\vec{a}$, but with a magnitude twice the magnitude of $\vec{a}$.

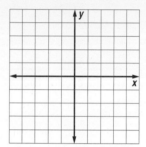

1. Write $\vec{a}$ and $\vec{b}$ in component form.

2. What do you notice about the components of $\vec{a}$ and $\vec{b}$?

3. Draw $\vec{b}$ so that its magnitude is three times that of $\vec{a}$. How do the components of $\vec{a}$ and $\vec{b}$ compare?

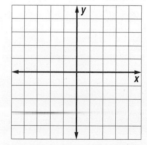

Make a Conjecture

4. Describe the vector magnitude and direction of a vector $\langle x, y \rangle$ after the components are multiplied by n.

Chapter 10 Circles
Teaching Notes and Overview

Geometry Activity Recording Sheet
Circumference Ratio
(p. 161 of this booklet)

Use With the activity on page 524 in Lesson 10-1 of the Student Edition.

Objective Discover pi as a ratio.

Materials
10 round objects
tape measure*
* = available in Overhead Manipulative Resources

Students can work in groups of three or four. Each member of the group can be responsible for gathering a few round objects. Remind students to measure each length to the nearest millimeter. Ask a member from each group to share the group's conjecture made in Exercise 3.

Answers
See Teacher Wraparound Edition page 524.

Geometry Activity Recording Sheet
Congruent Chords and Distance
(p. 162 of this booklet)

Use With the activity on page 538 in Lesson 10-3 of the Student Edition.

Objective Use patty paper to discover that congruent chords are equidistant from the center of a circle.

Materials
patty paper or tracing paper
compass*
scissors*
centimeter ruler*
* = available in Overhead Manipulative Resources

Students can work in pairs. Students first make a circle out of patty paper. By folding, bisecting, and measuring two chords, students discover that the chords are congruent and equidistant from the center of the circle.

Answers
See Teacher Wraparound Edition page 538.

Using Overhead Manipulatives
Locating the Center of a Circle
(p. 163 of this booklet)

Use With Lesson 10-3.

Objective Use constructions to find the center of a circle.

Materials
compass*
straightedge
transparency pens*
blank transparency
* = available in Overhead Manipulative Resources

This demonstration involves constructing and investigating the intersection of the perpendicular bisectors of two nonparallel chords of a circle. Before claiming point X as the center of the circle, you may wish to perform the construction with another pair of nonparallel chords.

In the extension, students discuss other methods to finding the center of a circle. Allow the representatives of each group to demonstrate their method on the overhead.

Answers
Answers appear on the teacher demonstration instructions on page 163.

© Glencoe/McGraw-Hill   Teaching Geometry with Manipulatives

Chapter 10 Teaching Notes and Overview

Mini-Project
More About Circles
(p. 164 of this booklet)

Use With Lesson 10-3.

Objective Use circles and paper folding to discover characteristics and properties of chords.

Materials
unlined paper
compass*
ruler*
scissors*
* = available in Overhead Manipulative Resources

For Exercises 1 and 2, students use paper folding to create two different regular polygons. In Exercises 3 and 4, students draw or fold the perpendicular bisectors of each chord to locate the centers of the circles. Students discover that they can use parallel and nonparallel chords to locate the center of a circle.

Answers

1. octagon 2. hexagon 3. See students' work.

4. Yes; the perpendicular bisectors are the same segment. Extend this segment until it becomes a diameter, then bisect it.

Geometry Activity Recording Sheet
Measure of Inscribed Angles
(p. 165 of this booklet)

Use With the activity on page 544 in Lesson 10-4 of the Student Edition.

Objective Investigate the measure of an angle inscribed in a circle.

Materials
compass*
protractor*
straightedge
* = available in Overhead Manipulative Resources

In this activity, students construct a circle and an inscribed angle. Students then draw two radii to form a central angle with the same endpoints as the inscribed angle. By measuring the angles and arc, students make a conjecture that suggests the Inscribed Angle Theorem.

Answers
See Teacher Wraparound Edition page 544.

Using Overhead Manipulatives
Investigating Inscribed Angles
(p. 166 of this booklet)

Use With Lesson 10-4.

Objective Investigate the measure of an angle inscribed in a circle.

Materials
compass*
protractor*
straightedge
transparency pens*
blank transparency
* = available in Overhead Manipulative Resources

This demonstration involves drawing a central angle and several inscribed angles that intercept the same arc to discover the relationship among these two types of angles. While students are examining circles of different sizes at their desks, encourage them to draw central angles of various sizes. Remind students to draw the circle large enough that the angles are easy to measure with a protractor.

Answers
Answers appear on the teacher demonstration instructions on page 166.

© Glencoe/McGraw-Hill

Chapter 10 Teaching Notes and Overview

Geometry Activity
Inscribed Angles
(pp. 167–169 of this booklet)

Use With Lesson 10-4.

Objective Define and measure inscribed and central angles of a circle.

Materials
classroom set of Geometry Activity worksheets
transparency master of Geometry Activity
blank transparencies

Display the transparency master on the overhead. Ask students to find how many degrees there are from one dot on the circle to the next. Use a blank transparency to draw a central angle on the circular grid. Ask students to find the measure of the central angle.

Students can work in pairs. As you review Exercise 5, select students to come to the overhead and draw the inscribed angles. You can also use a circle geoboard. However, the circle geoboard only has half as many dots as the circular grid, so the activity may need to be altered.

Answers
1a. 30   **1b.** 90   **1c.** 150   **1d.** 150
2a. 75, 75   **2b.** 45, 45   **2c.** 15, 15   **2d.** 15, 15
3. 30, 120, 150; They are equal.
4. They are congruent; answers may vary.
5. See students' angles.
6a. 45, 75, 60, 180   **6b.** 90, 90, 90, 90, 360
6c. 127.5, 127.5, 90, 82.5, 112.5, 540
6d. 120, 120, 120, 120, 120, 120, 720
7. Yes; sum of measures of angles = $(n-2)180$, where n = number of sides

Using Overhead Manipulatives
Constructing a Circle to Inscribe a Triangle
(p. 170 of this booklet)

Use With Lesson 10-5.

Objective Construct a circle so that a given triangle is inscribed in it.

Materials
compass*
straightedge
transparency pens*
blank transparencies
* = available in Overhead Manipulative Resources

This demonstration involves constructing a circle that inscribes a triangle. After drawing $\triangle RST$, you may wish to ask students for suggestions on how to construct a circle that will inscribe the triangle.

In this extension, students work in pairs to explore the types of polygons (regular, convex, and concave) that can be inscribed in a circle. You may wish to select a few students to demonstrate at the overhead how to inscribe a polygon, other than a triangle, in a circle.

Answers
Answers appear on the teacher demonstration instructions on page 170.

Using Overhead Manipulatives
Constructing Tangents
(pp. 171–172 of this booklet)

Use With Lesson 10-5.

Objective Construct a tangent to a circle through a point on or outside of the circle.

Chapter 10 Teaching Notes and Overview

Materials
compass*
straightedge
transparency pens*
blank transparency
* = available in Overhead Manipulative Resources

This activity includes two demonstrations on constructing tangents to a circle. You may need to review with students how to construct perpendicular lines through a point on the line and through a point not on the line.
- Demonstration 1 involves constructing a tangent through a point on the circle. To verify that the line is tangent, you can have a student come to the overhead and use a protractor to measure the angle formed by $\overrightarrow{CD}$ and the tangent line.
- Demonstration 2 involves constructing a tangent through a point not on the circle.
- In this extension, students investigate the relationship between two tangents to a circle from a common exterior point.

Answers
Answers appear on the teacher demonstration instructions on pages 171–172.

Using Overhead Manipulatives
Inscribing a Circle in a Triangle
(p. 173 of this booklet)

Use With Lesson 10-5.

Objective Inscribe a circle in a given triangle.

Materials
compass*
straightedge
transparency pens*
blank transparencies
* = available in Overhead Manipulative Resources

This demonstration involves constructing a circle that is inscribed in a triangle. You may wish to ask two students to come to the overhead and construct the angle bisector of $\angle L$ and $\angle N$.

In this extension, students prove that a circle can be inscribed in any triangle. Before determining how to prove this, you may encourage students to complete the construction at their desk with a triangle of any size.

Answers
Answers appear on the teacher demonstration instructions on page 173.

Geometry Activity Recording Sheet
Inscribed and Circumscribed Triangles
(pp. 174–175 of this booklet)

Use With Lessons 10-4 and 10-5 as a follow-up activity. This corresponds to the activity on pages 559–560 in the Student Edition.

Objective Construct circles inscribed in triangles and construct circles circumscribed about triangles.

Materials
compass*
straightedge
* = available in Overhead Manipulative Resources

In this activity, students learn how to construct a circle inscribed in a triangle and a circle circumscribed about a triangle. You may wish to select three students to perform the constructions in Exercises 1–3 at the chalkboard or overhead. Refer to these constructions and those in Activities 1 and 2 as you discuss the answers to Exercises 4–10.

Answers
See Teacher Wraparound Edition page 560.

© Glencoe/McGraw-Hill   Teaching Geometry with Manipulatives

NAME _____ DATE _____ PERIOD _____

Geometry Activity Recording Sheet

(Use with the activity on page 524 in Lesson 10–1 in the Student Edition.)

Circumference Ratio

Materials
10 round objects
tape measure

Gather Data and Analyze

1. Measure the circumference and diameter of each object using a millimeter measuring tape. Record the measures in the table below.

Object	C	d	$\frac{C}{d}$
1			
2			
3			
4			
5			
6			
7			
8			
9			
10			

2. Compute the value of $\frac{C}{d}$ to the nearest hundredth for each object. Record the result in the fourth column of the table above.

Make a Conjecture

3. What seems to be the relationship between the circumference and the diameter of the circle?

© Glencoe/McGraw-Hill   Teaching Geometry with Manipulatives

NAME _____ DATE _____ PERIOD ____

Geometry Activity Recording Sheet

(Use with the activity on page 538 in Lesson 10-3 of the Student Edition.)

Congruent Chords and Distance

Materials
patty paper or tracing paper
compass
scissors
centimeter ruler

Analyze

1. What is the relationship between $\overline{SU}$ and $\overline{VT}$? $\overline{SX}$ and $\overline{WY}$?

2. Use a centimeter ruler to measure $\overline{VT}$, $\overline{WY}$, $\overline{SU}$, and $\overline{SX}$. Record your measures in the table below. What do you find?

Segment	Measure
$\overline{VT}$	
$\overline{WY}$	
$\overline{SU}$	
$\overline{SX}$	

3. **Make a conjecture** about the distance that two chords are from the center when they are congruent.

Using Overhead Manipulatives
(Use with Lesson 10-3.)

Locating the Center of a Circle

Objective Use construction to find the center of a circle.

Materials
- compass*
- straightedge
- transparency pens*
- blank transparency

* = available in Overhead Manipulative Resource

Demonstration
Locate the Center of a Circle
- Use the compass to draw any circle on a blank transparency. Then, draw any two nonparallel chords. Label the chords $\overline{TU}$ and $\overline{VW}$.

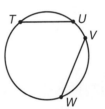

- Construct the perpendicular bisector of each chord. Call them ℓ and m. Label the intersection of ℓ and m point X.

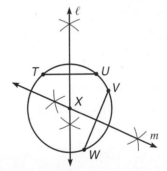

- Ask students to explain why we can claim that X is the center of the circle. **ℓ and m contain diameters of the circle since the perpendicular bisector of a chord must be a diameter. The only point two diameters of a circle share is the center of the circle.**

Extension
More Ways to Locate the Center of a Circle
- Divide students into groups of three or four. Have students work together to discover as many ways of finding the center of a circle as they can. Encourage students to use paper folding, construction, or another technique. Ask a representative from each group to demonstrate and justify one of their methods for finding the center.

NAME _____ DATE _____ PERIOD ____

Mini-Project

(Use with Lesson 10-3.)

More About Circles

1. Complete each of the following steps.

   a. Draw a circle with a 4-inch radius and cut it out.

   b. Fold the circle in half.

   c. Fold the circle in half again.

   d. Fold the circle in half one more time.

   e. Unfold the circle and draw a chord between each of the adjacent endpoints created by the folds.

   What figure have you drawn?

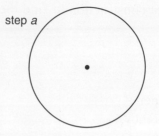
step a

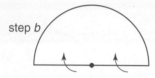
step b

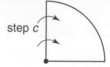
step c

step d

2. Complete each of the following steps.

   a. Draw another 4-inch radius circle and cut it out.

   b. Fold the circle in half.

   c. Fold the two sides so they meet in the middle as shown.

   d. Unfold the circle and draw a chord between each of the adjacent endpoints created by the folds.

   What figure have you drawn?

step a

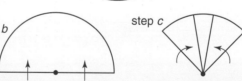
step b   step c

3. In a circle, the perpendicular bisector of any chord passes through the center of the circle. If you choose any two chords that are not parallel, then the intersection of their perpendicular bisectors must be the center of the circle. Use this fact to find the center of each of the following circles.

   a. 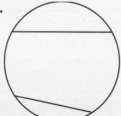   b. 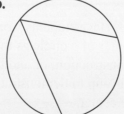   c.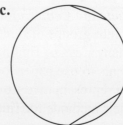

4. Draw a circle. Draw two chords that are parallel. Using those two chords, can you find the center of the circle?

NAME _____ DATE _____ PERIOD ____

Geometry Activity Recording Sheet

(Use with the activity on page 544 in Lesson 10-4 in the Student Edition.)

Measure of Inscribed Angles

Materials
compass
protractor
straightedge

Model
Draw $\odot W$ in the space provided. Draw inscribed angle *XYZ*. Draw $\overline{WX}$ and $\overline{WZ}$.

Analyze

1. Measure $\angle XYZ$ and $\angle XWZ$.

2. Find $m\,\widehat{XZ}$ and compare it with $m\angle XYZ$.

3. **Make a conjecture** about the relationship of the measure of an inscribed angle and the measure of its intercepted arc.

Geometry—Chapter 10

© Glencoe/McGraw-Hill   165   *Teaching Geometry with Manipulatives*

Using Overhead Manipulatives
(Use with Lesson 10-4.)

Investigating Inscribed Angles

Objective Investigate the measure of an angle inscribed in a circle.

Materials
- compass*
- protractor*
- straightedge
- transparency pens*
- blank transparency

* = available in Overhead Manipulative Resources

Demonstration
Investigate Inscribed Angles

- Draw a large circle C on a blank transparency. Using a straightedge, draw a central angle. Find the measure of the central angle with a protractor and record the measure on the transparency.

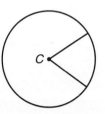

$m\angle C = 68$

- Draw several inscribed angles that intercept the same arc as the central angle. Have students measure each inscribed angle and record these measures on the transparency.

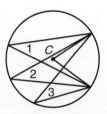

$m\angle C = 68$
$m\angle 1 = 34$
$m\angle 2 = 34$
$m\angle 3 = 34$

- Have students work in pairs to repeat the investigation with a few more circles of different sizes. Ask students how the measure of each inscribed angle relates to the measure of the central angle. **If an angle is inscribed in a circle, then its measure equals one-half the measure of its intercepted arc.**

© Glencoe/McGraw-Hill   Teaching Geometry with Manipulatives

NAME _____ DATE _____ PERIOD ____

Geometry Activity
(Use with Lesson 10-4.)

Inscribed Angles

NAME _____ DATE _____ PERIOD _____

Geometry Activity
(Use with Lesson 10-4.)

Inscribed Angles

1. Discuss the measures of the central angles in the figures below with your partner.

a. 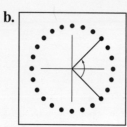 b. 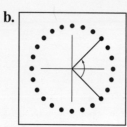 c. 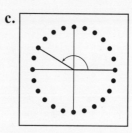 d.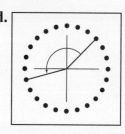

_____   _____   _____   _____

2. Connect the endpoints of the intercepted arcs in the figures of Exercise 1, and record the measures of the other two angles.

a. _____   _____   b. _____   _____

c. _____   _____   d. _____   _____

3. Refer to the figure below and complete the following:

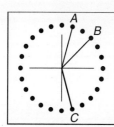

measure of $\widehat{AB}$ = _____

measure of $\widehat{BC}$ = _____

measure of $\widehat{AC}$ = _____

How do the arc measures compare to the measures of their respective central angles?

_____

4. Model two congruent inscribed angles on the circular grid.

What is true about their intercepted arcs? _____

Compare your model with your partner's design. Are they the same?

_____

List possible conclusions from your models. _____

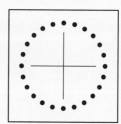

_____

© Glencoe/McGraw-Hill

Geometry Activity

5. On the circular grids draw inscribed angles measuring 45°, 30°, 60°, 90°, and 120°.

      45°   30°   60°   90°   120°

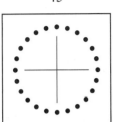 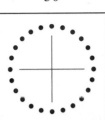 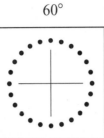 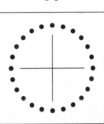 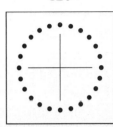

6. Record the measures of the angles in each figure below.

  a.   **b.**   **c.**   **d.**

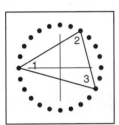 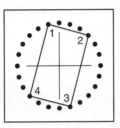 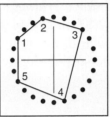 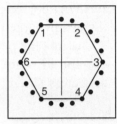

$m\angle 1$ _____ $m\angle 1$ _____ $m\angle 1$ _____ $m\angle 1$ _____
$m\angle 2$ _____ $m\angle 2$ _____ $m\angle 2$ _____ $m\angle 2$ _____
$m\angle 3$ _____ $m\angle 3$ _____ $m\angle 3$ _____ $m\angle 3$ _____
Total = _____ $m\angle 4$ _____ $m\angle 4$ _____ $m\angle 4$ _____
                       Total = _____ $m\angle 5$ _____ $m\angle 5$ _____
                                        Total = _____ $m\angle 6$ _____
                                                                          Total = _____

7. Can you find a pattern from the totals in Exercise 6?

_____

What conclusions can you make about the sum of the measures of the interior angles of an inscribed polygon?

_____

Using Overhead Manipulatives
(Use with Lesson 10-5.)

Constructing a Circle to Inscribe a Triangle

Objective Construct a circle so that a given triangle is inscribed in it.

Materials
- compass*
- straightedge
- transparency pens*
- blank transparencies

* = available in Overhead Manipulative Resources

Demonstration
Construct a Circle to Inscribe a Triangle
- Draw any large triangle *RST* on a blank transparency. Construct the perpendicular bisectors for two sides, $\overline{RS}$ and $\overline{ST}$. Label the intersection of the perpendicular bisectors *C*.

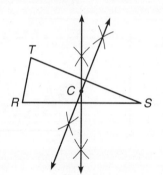

- Using *C* as a center, and *CR* as the measure of the radius, draw ⊙*C*. Then △*RST* is inscribed in ⊙*C*.

- Ask students "What kind of segments in a circle do the sides of △*RST* represent?" **chords** Ask students to explain why the construction works.
  The perpendicular bisector of a chord of a circle contains a diameter of the circle, so the perpendicular bisectors of $\overline{RS}$ and $\overline{ST}$ contain diameters of a circle in which $\overline{RS}$ and $\overline{ST}$ are chords.

Extension
Investigate Polygons that can be Inscribed in a Circle
- Explain that not all polygons can be inscribed in a circle. Have students work in pairs to investigate which polygons cannot be inscribed in a circle. Suggest that students organize their investigation by making a table listing all of the different types of polygons with up to eight sides that they can think of. Include regular, convex, and concave polygons. Then use the table to record which types of polygons can or cannot be inscribed in a circle. **Concave polygons cannot be inscribed in a circle.**

Using Overhead Manipulatives
(Use with Lesson 10-5.)

Constructing Tangents

Objective Construct a tangent to a circle through a point on or outside of the circle.

Materials
- compass*
- straightedge
- transparency pens*
- blank transparencies

* = available in Overhead Manipulative Resources

Demonstration 1
Construct a Tangent through a Point on a Circle
- Use a compass to draw a circle C. Choose a point on the circle and label it D. Draw $\overrightarrow{CD}$.

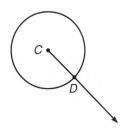

- Construct line ℓ through D and perpendicular to $\overrightarrow{CD}$. You may wish to refer students to the construction of a perpendicular line through a point on the line in Lesson 1-5 of *Geometry*.

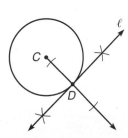

- Ask students to explain why ℓ is tangent to $\odot C$ at D.
  ℓ is perpendicular to radius $\overline{CD}$ at its endpoint, D, on the circle, so line ℓ is tangent to $\odot C$.

Demonstration 2
Construct a Tangent through a Point Not on a Circle
- Use a compass to draw a circle S. Choose a point outside of the circle and label it T. Draw $\overline{TS}$.

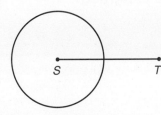

© Glencoe/McGraw-Hill   Teaching Geometry with Manipulatives

Using Overhead Manipulatives

- Construct the perpendicular bisector of $\overline{TS}$. Call this line ℓ. Label the intersection of ℓ and $\overline{TS}$ point R.

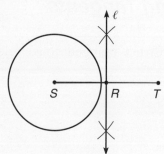

- Using R as the center, draw a circle with radius measuring RS. Call U and V the intersection points of the two circles.

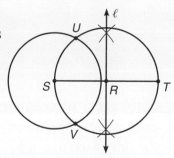

- Draw $\overleftrightarrow{TU}$ Then $\overleftrightarrow{TU}$ is tangent to $\odot S$.

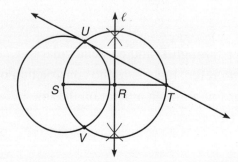

- Point out that $\overleftrightarrow{TU}$ is tangent to $\odot S$ if $\overleftrightarrow{TU} \perp \overline{US}$. Ask, "How do you know that $\angle SUT$ is a right angle?" **It is inscribed in a semicircle.**

Extension
Construct Two Tangents from an Exterior Point
- Have students work in pairs to construct the two tangents to a circle from an exterior point. Have them repeat the construction and measure each of the segments from the exterior point to the point of tangency. Ask students to make a conjecture about the measures of two tangent segments from an exterior point. **The two tangent segments from an exterior point to a circle are congruent.**

Using Overhead Manipulatives
(Use with Lesson 10-5.)

Inscribing a Circle in a Triangle

Objective Inscribe a circle in a given triangle.

Materials
- compass*
- straightedge
- transparency pens*
- blank transparencies

* = available in Overhead Manipulative Resources

Demonstration
Inscribe a Circle in a Triangle
- Draw a large triangle on a blank transparency. Label it $\triangle LMN$. Construct the angle bisector of $\angle L$ and $\angle N$. You may wish to refer students to the construction of an angle bisector in Lesson 1-4 of *Glencoe Geometry*. Extend the bisectors to intersect at point P.

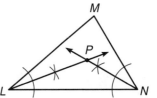

- Construct a line from P perpendicular to $\overline{LN}$. Label the intersection of the perpendicular line and $\overline{LN}$ as point Q.

- Set the compass length equal to PQ and draw $\odot P$.

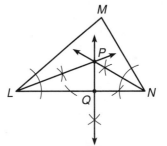

- Ask students to explain why the construction works. **Every point on the bisector of an angle is equidistant from the sides of the angle, so the intersection of the bisectors is equidistant from all three sides of the triangle. Therefore, a circle drawn with the intersection point as the center and the distance from the point to one of the sides as the measure of the radius will be tangent to all three sides.**

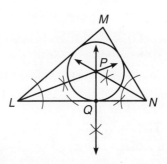

Extension
Inscribe a Circle in Any Triangle
- Have students work in pairs to prove that a circle can be inscribed in any triangle.

© Glencoe/McGraw-Hill   173   Teaching Geometry with Manipulatives

NAME _____ DATE _____ PERIOD ____

Geometry Activity Recording Sheet

(Use with the Lesson 10-4 and 10-5 Follow-Up Activity on pages 559–560 in the Student Edition.)

Inscribed and Circumscribed Triangles

Materials
compass
straightedge

Model

1. Draw an obtuse triangle and inscribe a circle in it.

2. Draw a right triangle and circumscribe a circle about it.

3. Draw a circle of any size and circumscribe an equilateral triangle about it.

© Glencoe/McGraw-Hill   174   Teaching Geometry with Manipulatives

Analyze
Refer to Activity 1.

4. Why do you only have to construct the perpendicular to one side of the triangle?

5. How can you use the Incenter Theorem to explain why this construction is valid?

Refer to Activity 2.

6. Why do you only have to measure the distance from the circumcenter to any vertex?

7. How can you use the Circumcenter Theorem to explain why this construction is valid?

Refer to Activity 3.

8. What is the measure of each of six congruent arcs?

9. Write a convincing argument as to why the lines constructed in Step 3 form an equilateral triangle.

10. Why do you think the terms incenter and circumcenter are good choices for the points they define?

Chapter 11 Areas of Polygons and Circles
Teaching Notes and Overview

Geometry Activity Recording Sheet
Area of a Parallelogram
(p. 180 of this booklet)

Use With the activity on page 595 in Lesson 11-1 of the Student Edition.

Objective Discover how the area formula for a parallelogram is similar to the area formula for a rectangle.

Materials
grid paper
straightedge

In this activity, students draw a parallelogram on a piece of grid paper. Then by folding the parallelogram to form a rectangle, students discover that the base of the parallelogram is half the length of the rectangle, and the altitude of the parallelogram is the same as the width of the rectangle. Through this discovery, students can make a conjecture about the formula for the area of a parallelogram.

Answers
See Teacher Wraparound Edition page 595.

Geometry Activity Recording Sheet
Area of a Triangle
(p. 181 of this booklet)

Use With the activity on page 601 in Lesson 11-2 of the Student Edition.

Objective Derive a formula for the area of a triangle.

Materials
grid paper
straightedge
scissors*

* = available in Overhead Manipulative Resources

Students begin this activity by drawing a triangle on a piece of grid paper. Then students draw lines that are perpendicular or parallel to the base of this triangle. Ask students what the triangle and rectangle have in common. By cutting out the rectangle and then the triangle, students discover that the two smaller triangles are the same size as $\triangle ABC$. This leads students to the fact that the area of a triangle is half the area of a rectangle with the same base and height.

Answers
See Teacher Wraparound Edition page 601.

Using Overhead Manipulatives
Investigating the Area of a Trapezoid
(pp. 182–183 of this booklet)

Use With Lesson 11-2.

Objective Derive the formula for the area of a trapezoid.

Materials
straightedge
centimeter grid transparency*
lined paper transparency*
transparency pens*
centimeter grid paper
scissors*
tape

* = available in Overhead Manipulative Resources

In this demonstration, students discover the similarities and differences between the area formulas for a parallelogram and a trapezoid.

© Glencoe/McGraw-Hill   Teaching Geometry with Manipulatives

Chapter 11 Teaching Notes and Overview

Through the students' knowledge of the formula for the area of a parallelogram, they can develop the formula for the area of a trapezoid.

In the extension, students repeat the activity with two different trapezoids. Ask students whether this activity will work with a trapezoid of any size. Have them explain their reasoning.

Answers
Answers appear on the teacher demonstration instructions on pages 182–183.

Using Overhead Manipulatives
Constructing a Regular Hexagon
(p. 184 of this booklet)

Use With Lesson 11-3.

Objective Construct a regular hexagon.

Materials
straightedge
compass*
transparency pens*
blank transparency
* = available in Overhead Manipulative Resources

This demonstration involves using a compass and straightedge to construct a regular hexagon. Since students will use hexagon *HIJKLM* for the extension, you may wish to have the students complete the construction of the hexagon at their desks while you complete the construction at the overhead.

In the extension, students work in pairs to construct an apothem of hexagon *HIJKLM*.

Answers
Answers appear on the teacher demonstration instructions on page 184.

Geometry Activity Recording Sheet
Area of a Circle
(p. 185 of this booklet)

Use With the activity on page 611 in Lesson 11-3 of the Student Edition.

Objective Discover that the areas of polygons approach the area of a circle.

Materials
none

In this activity, students examine the areas of polygons as the number of sides increases. By finding the area of each polygon, students discover that the areas of polygons approaches the area of a circle. Ask students to approximate the area of a 30-sided polygon or a 40-sided polygon.

Answers
See Teacher Wraparound Edition page 611.

Geometry Activity
Area of a Regular Polygon
(pp. 186–188 of this booklet)

Use With Lesson 11-3.

Objective Derive the area formula for a regular polygon.

Materials
classroom set of Geometry Activity worksheets
scissors*
straightedge
* = available in Overhead Manipulative Resources

In this activity, students use a regular hexagon inscribed in a circle to derive the formula for the area of a regular polygon. Throughout the activity, you may wish to assess the students' progress by checking their answers to Exercises 7, 10, and 18. The formula for the area of a regular polygon is derived by examining the area of the hexagon in relation to the area of the triangles that compose the hexagon. Once students complete the activity, you may wish to have them repeat the process with another regular polygon.

Answers

1–5. See students' work.

6. radii

7. $p = AB + BC + CD + DE + EF + FA$

8. 6

9. They are congruent.

10. The area of the hexagon equals the sum of the areas of the triangles.
    area of $ABCDE$ = area of $\triangle AOB$ + area of $\triangle BOC$ + area of $\triangle COD$ + area of $\triangle DOE$ area of $\triangle EOF$ + area of $\triangle FOA$

11. See students' work.

12. They are perpendicular.

13. the height

14. $\frac{1}{2}aDE$

15. yes

16. They are equal.

17a. $\frac{1}{2}aCD$

17b. $\frac{1}{2}aBC$

17c. $\frac{1}{2}aAB$

17d. $\frac{1}{2}aAF$

17e. $\frac{1}{2}aEF$

18. area of hexagon $ABCDEF = \frac{1}{2}aDE + \frac{1}{2}aCD + \frac{1}{2}aBC + \frac{1}{2}aAB + \frac{1}{2}aAF + \frac{1}{2}aEF$

19. area of hexagon $ABCDEF = \frac{1}{2}a(DE + CD + BC + AB + AF + EF)$

20. area of hexagon $ABCDEF = \frac{1}{2}ap$

21. Sample answer: A regular polygon can be separated into congruent nonoverlapping triangles. The sum of the areas of the triangles is the area of the polygon. The area of one triangle is $\frac{1}{2}$ times the product of the base and the height. The base of the triangle is the length of one side of the polygon and the height is the apothem of the polygon. This can be rewritten as $\frac{1}{2}sa$. The area of a regular polygon is $\frac{1}{2}Pa$ where P is the perimeter of the polygon or the sum of the sides, s, of each triangle.

Mini-Project

Areas of Circular Regions
(p. 189 of this booklet)

Use With Lesson 11-3.

Objective Find the area of the space that the dog can reach while on a 12-foot chain.

Materials
compass

For this activity, students should work in pairs. Students should first use a compass to draw the circular region in each figure with the radius defined by the segment labeled "12 ft." Then students will use the formula for the area of a circle to calculate the space in each figure that Rover can reach while on the 12-foot chain. Remind students that each circular region, except in Exercise 1, is only a portion of a circle.

Answers

1. 452.4 ft²
2. 226.2 ft²
3. 339.3 ft²
4. 389.6 ft²

NAME _____ DATE _____ PERIOD ____

Geometry Activity Recording Sheet

(Use with the activity on page 595 in Lesson 11-1 in the Student Edition.)

Area of a Parallelogram

Materials
grid paper
straightedge

Analyze

1. What is the area of the rectangle?

2. How many rectangles form the parallelogram?

3. What is the area of the parallelogram?

4. How do the base and altitude of the parallelogram relate to the length and width of the rectangle?

5. **Make a conjecture** Use what you observed to write a formula for the area of a parallelogram.

NAME _____ DATE _____ PERIOD ____

Geometry Activity Recording Sheet

(Use with the Lesson 11-2 Follow-Up Activity on page 601 in the Student Edition.)

Area of a Triangle

Materials
grid paper
straightedge
scissors

Analyze

1. What do you observe about the two smaller triangles and $\triangle ABC$?

2. What fraction of rectangle $ACDE$ is $\triangle ABC$?

3. Derive a formula that could be used to find the area of $\triangle ABC$.

Using Overhead Manipulatives
(Use with Lesson 11-2.)

Investigating the Area of a Trapezoid

Objective Find the area of a trapezoid, and develop a formula for the area of a trapezoid.

Materials
- straightedge
- centimeter grid transparency*
- lined paper transparency*
- transparency pens*
- centimeter grid paper
- scissors*
- tape

* = available in Overhead Manipulative Resources

Demonstration
Investigate the Area of a Trapezoid

- Review the definition of a trapezoid. Ask students how a parallelogram and a trapezoid are the same and how they are different. **They are both quadrilaterals, but a parallelogram has two pairs of sides parallel and a trapezoid has only one pair of sides parallel.**

- Ask students if the formula for the area of a parallelogram can be used to find the area of a trapezoid. Explain. **No, the bases of a trapezoid are not the same measure.**

- On the centimeter grid transparency, draw a trapezoid with bases 14 centimeters and 18 centimeters long and with a height of 8 centimeters. Label the bases and height as shown. Record the measures of the bases and the height in a chart on the lined paper transparency.

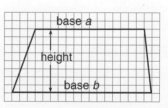

- Draw an identical trapezoid on centimeter grid paper. Cut out the trapezoid. Show students that it is the same size and shape as the trapezoid drawn on the transparency.

- Fold the paper trapezoid so that the bases align. Unfold and cut the paper trapezoid along the fold.

- Place the pieces together to form a parallelogram. Tape the pieces together and place on the centimeter grid transparency.

© Glencoe/McGraw-Hill   182   Teaching Geometry with Manipulatives

- Ask students to find the length of the base of the parallelogram. **32 cm** Ask them how it compares with the measures of the bases of the trapezoid. **It is the sum of the measures of the bases of the trapezoid.**

- Ask students to find the height of the parallelogram. **4 cm** Ask them how it compares with the height of the trapezoid. **It is half of the height of the trapezoid.**

- Ask students to find the area of the parallelogram. **128 cm²**

- Ask students what the area of the trapezoid is. **It is the same as the area of the parallelogram, 128 cm².**

- Ask students to write a formula for the area of any trapezoid. $A = \frac{1}{2}h(b_1 + b_2)$

Extension
Use the Area Formula for a Trapezoid
- Have students repeat the activity to find the area of each trapezoid described below.
    - bases: 13 centimeters and 17 centimeters; height: 16 centimeters **240 cm²**
    - bases: 5 centimeters and 15 centimeters; height: 14 centimeters **140 cm²**

Using Overhead Manipulatives
(Use with Lesson 11-3.)

Constructing a Regular Hexagon

Objective Construct a regular hexagon.

Materials
- straightedge
- compass*
- transparency pens*
- blank transparency

* = available in Overhead Manipulative Resources

Demonstration
Construct a Regular Hexagon
- Draw a circle with the compass. Choose a point on the circle and label it *H*.

- With the same compass setting, place the compass on point *H*. Draw a small arc that intersects the circle. Label the point of intersection with the circle point *I*. Place the compass on point *I*. Draw another small arc that intersects the circle. Label that intersection point *J*. Continue this process, labeling points *K*, *L*, and *M*, until you come back to point *H*.

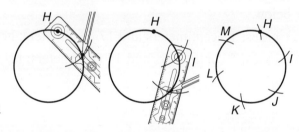

- Use a straightedge to connect the points *H*, *I*, *J*, *K*, *L*, and *M* in order. Ask students what is true about all of the segments. **They are congruent.**

- Ask students if this is a regular hexagon. **yes**

Extension
Construct an Apothem of a Regular Hexagon
- Explain that an apothem is a segment drawn from the center of a regular polygon perpendicular to a side of the polygon. Ask students to work in pairs to construct an apothem of hexagon *HIJKLM*. **Construct a line through *C* perpendicular to $\overline{HI}$.**

© Glencoe/McGraw-Hill   Teaching Geometry with Manipulatives

NAME _____ DATE _____ PERIOD _____

Geometry Activity Recording Sheet

(Use with the activity on page 611 in Lesson 11-3 in the Student Edition.)

Area of a Circle

Materials
none

Collect Data

1. Complete the following table. Round to the nearest hundredth.

Number of Sides	3	5	8	10	20	50
Measure of a Side	1.73r	1.18r	0.77r	0.62r	0.31r	0.126r
Measure of Apothem	0.5r	0.81r	0.92r	0.95r	0.99r	0.998r
Area						

Analyze the Data

2. What happens to the appearance of the polygon as the number of sides increases?

3. What happens to the areas as the number of sides increases?

4. **Make a conjecture** about the formula for the area of a circle.

© Glencoe/McGraw-Hill   Teaching Geometry with Manipulatives

NAME _____ DATE _____ PERIOD ____

Geometry Activity
(Use with Lesson 11-3.)

Area of a Regular Polygon

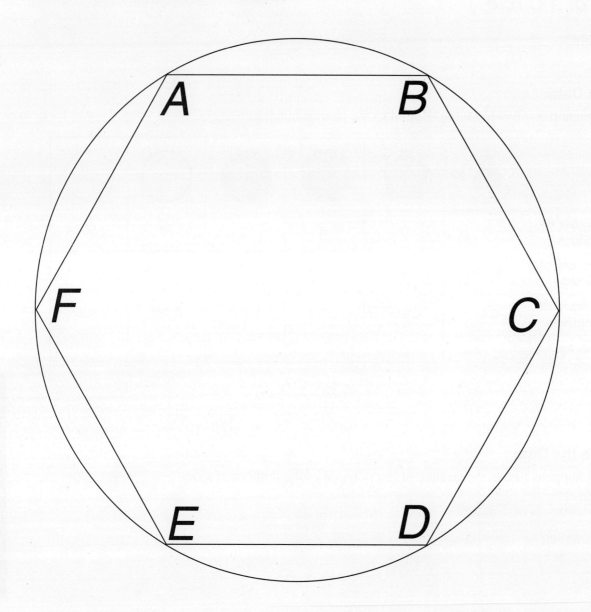

© Glencoe/McGraw-Hill   Teaching Geometry with Manipulatives

Geometry Activity

1. Cut out the circle circumscribed around hexagon ABCDEF.

2. Fold $\overline{AB}$ on $\overline{ED}$ and crease through F and C.

3. Open the circle and fold $\overline{BC}$ on $\overline{FE}$ and crease through A and D.

4. Open the circle and fold $\overline{CD}$ on $\overline{AF}$ and crease through B and E.

5. Draw $\overline{AD}, \overline{BE}$, and $\overline{CF}$ formed by the folds. Call the point where all the segments intersect, point O.

6. Consider circle O. What can you call $\overline{OA}, \overline{OB}, \overline{OC}, \overline{OD}, \overline{OE}$, and $\overline{OF}$?

7. Write an equation for the perimeter of hexagon ABCDEF.

8. How many triangles are formed by all the segments and all the sides of hexagon ABCDEF?

9. How are the triangles related?

10. How is the area of the hexagon related to the areas of the triangles? Write an equation to represent this.

11. Fold $\overline{ED}$ on itself and crease so that the crease passes through O. Label the intersection of $\overline{ED}$ and the crease point L. Let $\overline{OL}$ represent the apothem, a, of the polygon.

12. How are a and $\overline{ED}$ related?

13. Consider △EOD. What is a in △EOD?

Geometry Activity

14. What is the area of △EOD?

15. Do all six triangles have an apothem?

16. How are the apothems of all six triangles related?

17. Find the area of each triangle. (Hint: Let *a* represent the height of each triangle.)

   a. △DOC   b. △COB   c. △BOA

   d. △AOF   e. △FOE

18. Refer back to the equation you wrote in Exercise 10. Use substitution and the equations from Exercises 14 and 17 to write a new equation for the area of hexagon ABCDEF.

19. Use the distributive property and factor out the common factor of the equation in Exercise 18.

20. Refer back to the equation you wrote in Exercise 7. Use substitution to write a simplified equation for the area of hexagon ABCDEF.

21. Write a summary as to how you can use this method to prove that $A = \frac{1}{2}Pa$ is the area formula for any regular polygon.

Mini-Project

(Use with Lesson 11-3.)

Areas of Circular Regions

Robin is going to fix a chain to tie up his dog Rover. There are several places in the yard that Robin can attach the end of the chain. For each of the following, use a compass to draw the space that Rover can reach while on the end of a 12-foot chain. Then find the area. Round to the nearest tenth.

1. Rover's chain is attached to a stake in the middle of the yard.

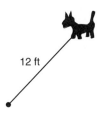

2. Rover's chain is attached to a long wall.

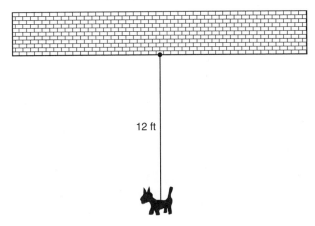

3. Rover's chain is attached to the corner byof the house.

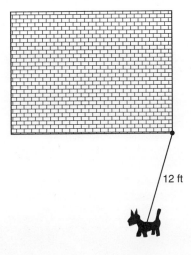

4. Rover's chain is attached to a 4-foot by 18-foot rectangular shed.

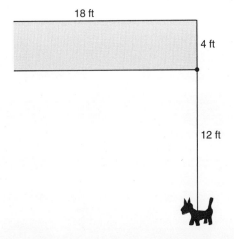

© Glencoe/McGraw-Hill

189

Teaching Geometry with Manipulatives

Chapter 12 Surface Area
Teaching Notes and Overview

Using Overhead Manipulatives
Drawing a Rectangular Solid
(pp. 194–195 of this booklet)

Use With Lesson 12-2.

Objective Draw a rectangular solid on isometric dot paper.

Materials
isometric dot paper transparency*
straightedge
transparency pens*
isometric dot paper for students
* = available in Overhead Manipulative Resources

This demonstration involves creating a rectangular prism on isometric dot paper. Remind students that a figure drawn on isometric dot paper represents a corner view of a three-dimensional figure. Once you complete the drawing on the overhead, you may wish to have students draw their own rectangular prism with different dimensions on isometric dot paper.

The extension explores how to draw prisms with a triangular base and with a trapezoidal base. With each of these figures, explain to students that it is not possible to determine the lengths of each side of the figure simply by counting the dots.

Answers
Answers appear on the teacher demonstration instructions on pages 194–195.

Geometry Activity
Surface Areas of Cylinders and Cones
(pp. 196–198 of this booklet)

Use With Lesson 12-6.

Objective Find the lateral area and surface area of a right cylinder and a right circular cone.

Materials
classroom set of Geometry Activity worksheets
transparency master of Geometry Activity
soup can with label
scissors*
ruler*
* = available in Overhead Manipulative Resources

You may wish to complete the following demonstration before handing out the worksheet.

Demonstration:
Place the transparency master on the overhead. Cover the transparency, revealing only Figure 1. Cut a soup can label along a line perpendicular to the base of the can, and remove the label. Lay the paper label flat and identify its area. Have students find the lateral area, $L = $ base $\times$ height, where base = circumference of the can. Use the can to find the total area, $T = L + 2$(area of circle). Uncover Figure 2 on the transparency master and discuss how this relates to the soup can example.

Have students work in groups of two or three to complete the activity. Before students build the cones, ask them how the radius of the sector compares with the slant height of the cone. When students have completed the activity, review the answers with the class.

Answers
1a. $L = 112\pi$, $T = 210\pi$
1b. $L = 240\pi$, $T = 1040\pi$
2a. πr^2
2b. $\dfrac{n}{360}\pi r^2$
2c. $2\pi r$

Chapter 12 Teaching Notes and Overview

3. All are 3 cm.
4. A: 6.75π, B: 2.25π, C: 6π, D: 3π
5. The lateral area of cone A is 3 times the lateral area of cone B. The lateral area of cone C is 2 times the lateral area of cone D.
6. ℓ: 3 cm, 3 cm, 3 cm, 3 cm; h: 2 cm, 2.9 cm, 2.3 cm, 2.8 cm; r: 2.25 cm, 0.75 cm, 2 cm, 1 cm
7. See students' work; for each cone in Exercise 6, check that $r^2 + h^2 = \ell^2$.
8. A: 6.75π, B: 2.25π, C: 6π, D: 3π

Mini-Project
Cone Patterns
(p. 199 of this booklet)

Use With Lesson 12-6.

Objective Draw, cut out, and find measurements of cone patterns.

Materials
construction paper
compass*
centimeter ruler*

* = available in Overhead Manipulative Resources

Students can work in groups of two or three for this activity. Using the given measurements, students draw and cut out each cone pattern. For each cone, students find the circumference of the base, height, slant height, lateral area, and surface area.

Answers
1. 5 cm
2. $\frac{3}{4}$
3. 31.4 cm
4. 23.56 cm
5. See students' work.
6. 7.5 cm

7. 23.56 cm
8. 5 cm
9. 3.3 cm
10. 58.9 cm^2
11. 103.1 cm^2
12. 8.0 cm; 75.4 cm^2; 4.5 cm
13. 10 cm; 157.1 cm^2; 8.7 cm

Geometry Activity Recording Sheet
Surface Area of a Sphere
(p. 200 of this booklet)

Use With the activity on page 672 in Lesson 12-7 of the Student Edition.

Objective Discover how the surface area of a sphere relates to the area of a great circle of the sphere.

Materials
polystyrene ball
scissors*
unlined paper
tape
glue

* = available in Overhead Manipulative Resources

For this activity, students can work in groups of two or three. Students first cut a polystyrene ball along a great circle. Then students trace the great circle and cut it into eight congruent pieces. Students then rearrange these pieces to form the pattern shown in the Student Edition. By taping this pattern to the polystyrene ball, students discover that this pattern, or the area of a great circle, is $\frac{1}{4}$ the surface area of a sphere.

Answers
See Teacher Wraparound Edition page 672.

© Glencoe/McGraw-Hill
Teaching Geometry with Manipulatives

Chapter 12 Teaching Notes and Overview

Geometry Activity Recording Sheet
Locus and Spheres
(p. 201 of this booklet)

Use With Lesson 12-7 as a follow-up activity. This corresponds to the activity on page 677 in the Student Edition.

Objective Examine spheres and the locus of points they represent.

Materials
straightedge

Students can work in pairs to complete this activity. Students investigate and identify the locus of points that represent a given set of criteria. You may wish to select a student to draw their figure for Exercise 1 on the chalkboard or overhead.

Answers
See Teacher Wraparound Edition page 677.

Using Overhead Manipulatives
Intersection of Loci
(pp. 202–203 of this booklet)

Use With Lesson 12-7.

Objective Construct a triangle given two sides and an altitude.

Materials
ruler*
compass*
transparency pens*
blank transparency
* = available in Overhead Manipulative Resources

This demonstration involves constructing a triangle that satisfies given conditions. Use this demonstration after discussing locus in the follow-up activity for Lesson 12-7. Once the four possible triangles are identifiable, you may wish to have students come to the overhead and outline them with colored transparency pens.

Answers
Answers appear on the teacher demonstration instructions on pages 202–203.

© Glencoe/McGraw-Hill   Teaching Geometry with Manipulatives

Using Overhead Manipulatives
(Use with Lesson 12-2.)

Drawing a Rectangular Solid

Objective Draw a rectangular solid on isometric dot paper.

Materials
- isometric dot paper transparency*
- straightedge
- transparency pens*
- isometric dot paper for students

* = available in Overhead Manipulative Resources

Demonstration
Draw a Rectangular Prism

- On the isometric dot paper transparency, use a colored transparency pen to draw a vertical line segment 4 units long. Draw diagonal lines representing a width of 2 units and a length of 3 units.

- Use a second colored pen to complete the base of the prism. (*Hint:* You may want to used dashed lines to represent the hidden parts of the figure.)

- Use a third colored pen to draw three vertical segments that are 4 units long from the vertices of the base.

- Using a fourth color pen, draw the segments that complete the top side of the figure. Use a black pen to label each segment with its length.

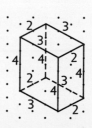

- Ask which of the faces represent the bases of the rectangular prism. **the rectangles that are 3 by 2 units**
- Ask whether the drawing would look different if you turned the figure over on its side. **Yes; it would be resting on a base that was 4 by 2 units, with a height of 3 units.** Draw the figure.

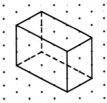

Extension
Draw a Triangular Prism and a Hexagonal Prism

- Ask how you could draw a prism with a base in the shape of a triangle. You may need to have students look back at the rectangular prism and ask, "If we started by drawing the base, in what direction would the segments for the height be drawn?" **Vertically; a sample triangular prism is shown.** Draw a triangular prism.

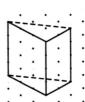

- Ask how you could draw a prism with a base in the shape of a trapezoid. Draw the figure at the right on the isometric dot paper transparency. Give students isometric dot paper. Have them copy the figure. Then ask them what two-dimensional shapes they will see from the top view, front view, left view, and right view. **top view: trapezoid; front view: rectangle; left view: rectangle; right view: rectangle**

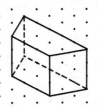

NAME _____ DATE _____ PERIOD ____

Geometry Activity
(Use with Lesson 12-6.)

Cylinders and Cones

1.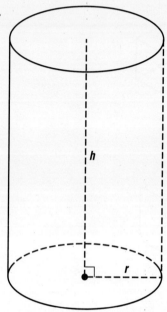

2.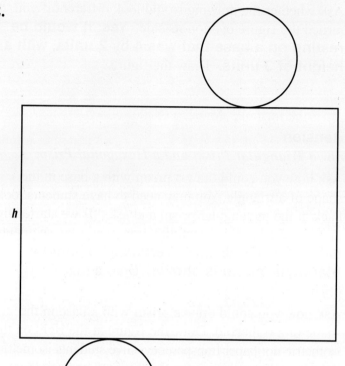

3.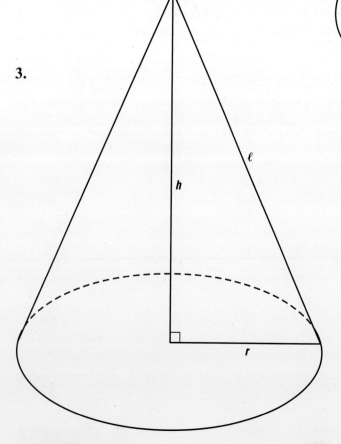

© Glencoe/McGraw-Hill

NAME _____ DATE _____ PERIOD ____

Geometry Activity

(Use with Lesson 12-6.)

Surface Areas of Cylinders and Cones

1. Find the lateral area L and total surface area T of each right cylinder.

   a. $L =$ _____   b. $L =$ _____

   $T =$ _____ $T =$ _____

   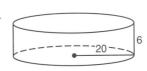

2. List the formulas for each of the following.

   a. area of a circle _____   b. area of a sector _____

   c. circumference of a circle _____

3. Measure the radius of the sectors on the next page.

   A _____   B _____   C _____   D _____

4. Find the lateral area of each sector from Exercise 3.

   A _____   B _____   C _____   D _____

5. How do the lateral areas compare for cones A and B? _____

   Cones C and D? _____

Construct four cones from the patterns and instructions on the next page.

6. Measure and record the following.

	Cone A	Cone B	Cone C	Cone D
Slant height (ℓ)				
height (h)				
radius of the base (r)				

7. Check the accuracy of your measurements using the Pythagorean Theorem.

   Example: $3^2 + 4^2 = 5^2$

   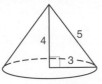

   A _____   B _____
   C _____   D _____

8. Using the slant height and radius from Exercise 6, find the lateral area of each cone. ($L = \pi r \ell$)

   A _____   B _____   C _____   D _____

© Glencoe/McGraw-Hill   Teaching Geometry with Manipulatives

Geometry Activity

Steps to Build a Cone
- Cut out the figures below.
- Tape edges (radii) together without overlapping.
- To measure the height of the cone, insert a pencil through the vertex of the cone, mark the length, then measure with a ruler.
- To measure the radius of the base, place the cone on a ruler and measure the widest distance (diameter).
  (radius = $\frac{1}{2}$ diameter)

A.
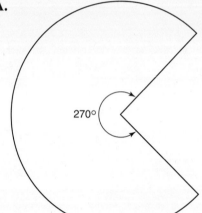

B.
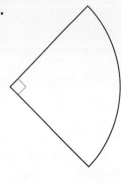

C.
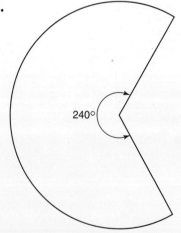

D.
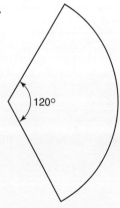

NAME _____ DATE _____ PERIOD ____

Mini-Project
(Use with Lesson 12-6.)

Cone Patterns

Using *AB* = 5 centimeters, draw a pattern shaped like the one at the right. It can be folded to make a cone.

1. Measure the radius of the circle to the nearest centimeter.

2. The pattern is what fraction of the complete circle?

3. What is the circumference of the complete circle?

4. How long is the circular arc that is the outside of the pattern?

5. Cut out the pattern and tape it together to form a cone.

6. Measure the diameter of the circular base of the cone.

7. What is the circumference of the base of the cone?

8. What is the slant height of the cone?

9. Use the Pythagorean Theorem to calculate the height of the cone. Use a decimal approximation. Check your calculation by measuring the height with a metric ruler.

10. Find the lateral area.

11. Find the total surface area.

Make a paper pattern for each cone with the given measurements. Then cut out the pattern and make the cone. Find the measurements to the nearest tenth.

12.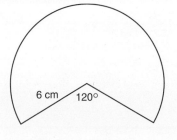

13.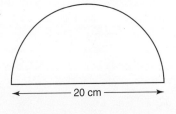

diameter of base = _____

lateral area = _____

height of cone = _____

diameter of base = _____

lateral area = _____

height of cone = _____

© Glencoe/McGraw-Hill   199   Teaching Geometry with Manipulatives

NAME _____ DATE _____ PERIOD ____

Geometry Activity Recording Sheet

(Use with the activity on page 672 in Lesson 12-7 of the Student Edition.)

Surface Area of a Sphere

Materials
polystyrene ball
scissors
unlined paper
tape
glue

Analyze

1. Approximately what fraction of the surface of the sphere is covered by the pattern?

2. What is the area of the pattern in terms of r, the radius of the sphere?

Make a Conjecture

3. Make a conjecture about the formula for the surface area of a sphere.

© Glencoe/McGraw-Hill — Teaching Geometry with Manipulatives

NAME _____ DATE _____ PERIOD ____

Geometry Activity Recording Sheet

(Use with the Lesson 12-7 Follow-Up Activity on page 677 in the Student Edition.)

Locus and Spheres

Materials
straightedge

Analyze

1. Draw a figure and describe the locus of points in space that are 5 units from each endpoint of a given segment that is 25 units long.

2. Are the two spheres congruent?

3. What are the radii and diameters of each sphere?

4. Find the distance between the two spheres.

5. What is the shape of the intersection of the spheres?

6. Can this be described as a locus of points in space or on a plane? Explain.

7. Describe the intersection as a locus.

8. **MINING** What is the locus of points that describe how particles will disperse in an explosion at ground level if the expected distance a particle could travel is 300 feet?

© Glencoe/McGraw-Hill   201   Teaching Geometry with Manipulatives

Using Overhead Manipulatives
(Use with Lesson 12-7.)

Intersection of Loci

Objective Construct a triangle given two sides and an altitude.

Materials
- ruler*
- compass*
- transparency pens*
- blank transparency

* = available in Overhead Manipulative Resources

Demonstration
Find the Intersection of Loci

- On a blank transparency, draw three segments to represent the measures RS, RT, and TU, such that $RS > RT > TU$.

  R•————————•S

  R•————•T T•————•U

- Tell students, "We want to construct $\triangle RST$ such that $\overline{RS}$ and $\overline{RT}$ are the sides of the triangle and $\overline{TU}$ is an altitude of the triangle."

- Draw $\odot R$ with radius of length RT. Draw $\overline{RS}$. Point out to students that vertex T of the triangle will lie RT units from R. Say, "The locus of all points RT units from R is a circle with center R and a radius of RT units." Ask students where vertex T will be located. **Vertex *T* will be somewhere on ⊙*R*.**

- Tell students, "Since vertex T is one endpoint of the altitude, endpoint U of altitude $\overline{TU}$ will lie on side $\overline{RS}$ or on the line containing $\overline{RS}$. Since $\overline{TU}$ is an altitude, then $\overline{TU} \perp \overleftrightarrow{RS}$. This means that endpoint T will lie on a line parallel to $\overleftrightarrow{RS}$, TU units from $\overleftrightarrow{RS}$." Ask students to describe the locus of all points meeting these conditions. **The locus of all points meeting these conditions are two lines on either side of $\overleftrightarrow{RS}$, parallel to $\overleftrightarrow{RS}$, at a distance of *TU* units.**

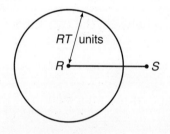

© Glencoe/McGraw-Hill

Teaching Geometry with Manipulatives

- Tell students, "Point T must satisfy the conditions for being a vertex and the conditions for being an endpoint of an altitude. Only four points satisfy these conditions."

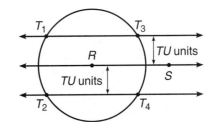

- Have four students describe the four possible ways to draw $\triangle RST$ and use a different colored transparency pen to draw each possible triangle.

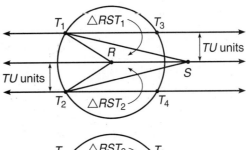

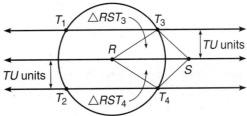

Chapter 13 Volume
Teaching Notes and Overview

Geometry Activity Recording Sheet
Volume of Rectangular Prism
(p. 206 of this booklet)

Use With the activity on page 688 in Lesson 13-1 of the Student Edition.

Objective Discover the formula for the volume of a rectangular prism.

Materials
centimeter cubes*
* = available in Overhead Manipulative Resources

For this activity, students can work in pairs. Students assemble centimeter cubes into a rectangular prism based on the orthogonal drawing. Then students compare the total number of cubes in the prism with the product of the length, width, and height of the prism. Students discover that the volume of a rectangular prism can be found by multiplying its length, width, and height.

Answers
See Teacher Wraparound Edition page 688.

Geometry Activity Recording Sheet
Investigating the Volume of a Pyramid
(p. 207 of this booklet)

Use With the activity on page 696 in Lesson 13-2 of the Student Edition.

Objective Derive the formula for the volume of a pyramid.

Materials
card stock
compass*
metric ruler*

scissors*
tape
rice
* = available in Overhead Manipulative Resources

For this activity, students can work in groups of three or four. Students draw, cut out, and assemble the rectangular prism and pyramid. Then students fill the rectangular prism with rice using the pyramid. (Instead of rice, you could also have students use beans or popcorn kernels.) Students then make the connection that the volume of a pyramid is $\frac{1}{3}$ the volume of a rectangular prism with the same base and height.

Answers
See Teacher Wraparound Edition page 696.

Mini-Project
Word Search
(p. 208 of this booklet)

Use With Lesson 13-4.

Objective Find and circle each word in the word search. Then find the meaning of each word and a page where it appears in your book.

For this activity, students can work in pairs. Ask students to identify how the meanings they find in a dictionary are different or like the meanings in their Geometry book.

Answers

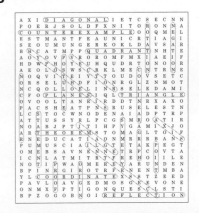

NAME _____ DATE _____ PERIOD _____

Geometry Activity Recording Sheet

(Use with the activity on page 688 in Lesson 13-1 of the Student Edition.)

Volume of a Rectangular Prism

Materials
centimeter cubes

Analyze

1. How many cubes make up the prism?

2. Find the product of the length, width, and height of the prism.

3. Compare the number of cubes to the product of the length, width, and height.

4. Repeat the activity with a prism of different dimensions.

5. **Make a conjecture** about the formula for the volume of a right rectangular prism.

NAME _____ DATE _____ PERIOD _____

Geometry Activity Recording Sheet

(Use with the activity on page 696 in Lesson 13-2 of the Student Edition.)

Investigating the Volume of a Pyramid

Materials
card stock
compass
metric ruler
scissors
tape
rice

Analyze

1. How many pyramids of rice did it take to fill the prism?

2. Compare the areas of the bases of the prism and pyramid.

3. Compare the heights of the prism and pyramid.

4. **Make a conjecture** about the formula for the volume of a pyramid.

NAME _____ DATE _____ PERIOD ____

Mini-Project
(Use with Lesson 13-4.)

Word Search

Each word in the list below can be found in your geometry book. Find and circle each of these words in the word-search puzzle. Then find the meaning of each word in a dictionary, and find a page in your book where each word appears.

AXIS	HYPOTHESIS	PLANES	ROTATION
COLLINEAR	IMAGE	POINT	STATEMENT
COORDINATE	INTERSECT	POSTULATE	SYMMETRY
COUNTEREXAMPLE	NEGATION	PROOF	THEOREM
DIAGONAL	NONCOPLANAR	QUADRANT	TRIANGLE
GRAPH	ORIGIN	REFLECTION	VECTOR

```
A X I D I A G O N A L I E T C S E C N N
P O E R J S O L D F X N I T O R O N N A
C O U N T E R E X A M P L E O O Q M E L
E S T M A N T F E A U N I C R T I A G I
S E O U M U N G E R K O K L D A V S A R
B H C A T M P P Q U A D R A N T N H T E
A O Y O V P V R O R O M F M X I A E I F
H D W P H O Y E U H Q U D R T O N O O R
A E O L O S W S C R K L M E C N T R N E
N O Q V I T E I Y T T O U D O V S E T C
O R S E L U H D F I O N R G L Z N M O T
N C Q O L L O E L I N R S E L E D A M I
C F O P L A N E S I Q L T R I A N G L E
O V O O L T A N R I R D D T N R X A X O
P A C S H E A T P N S R U S E L E S T N
L C S T O C W N O D E N A I A D P T R P
A T T U S S Y R L P C G S M R O A T I R
N O A B J P T I T I H P Y G A M I X J O
A R T H E O R E M S T O M A G L T O I F
R N E D U C A T I A O N M R R R B A N S
P U M U S C I A I L G T E T A K P E G T
O M E B S A V N E N N E T R P C O V T A
I C N L A T M I T R T F R E H O I I L E
N O T I P W A O M E R E Y A R U N D E N
B P I N R G I R O T R F R N E N T M B A
Y L C O O R D I N A T E X S S T Z E E D
P A V L O A V G E D M O S G E E V O N E
O N M X F P T I G O N Q U E S C L S T I
B P Z O G O B N O I R E F L E C T I O N
```

© Glencoe/McGraw-Hill   208   Teaching Geometry with Manipulatives